# CRC SERIES IN AGING

Editors-in-Chief: **Richard C. Adelman, Ph.D. and George S. Roth, Ph.D.**

VOLUMES AND VOLUME EDITORS

**HANDBOOK OF BIOCHEMISTRY IN AGING**
**James Florini, Ph.D.**
Syracuse University
Syracuse, New York

**HANDBOOK OF IMMUNOLOGY IN AGING**
**Marguerite M. B. Kay, M.D. and Takashi Makinodan, Ph.D.**
Geriatric Research Education and Clinical Center
V.A. Wadsworth Medical Center
Los Angeles, California

**SENESCENCE IN PLANTS**
**Kenneth V. Thimann, Ph.D.**
The Thimann Laboratories
University of California
Santa Cruz, California

**ALCOHOLISM AND AGING: ADVANCES IN RESEARCH**
**W. Gibson Wood, Ph.D.**
Geriatric Research Education and Clinical Center
V.A. Medical Center
St. Louis, Missouri
**Merrill F. Elias, Ph.D.**
University of Maine at Orono
Orono, Maine

**TESTING THE THEORIES OF AGING**
**Richard C. Adelman, Ph.D.**
University of Michigan
Ann Arbor, Michigan
**George S. Roth, Ph.D.**
Gerontology Research Center
National Institute on Aging
Baltimore City Hospitals
Baltimore, Maryland

**HANDBOOK OF PHYSIOLOGY IN AGING**
**Edward J. Masoro, Ph.D.**
University of Texas Health Science Center
San Antonio, Texas

**IMMUNOLOGICAL TECHNIQUES APPLIED TO AGING RESEARCH**
**William H. Adler, M.D. and**
**Albert A. Nordin, Ph.D.**
Gerontology Research Center
National Institute on Aging
Baltimore City Hospitals
Baltimore, Maryland

**CURRENT TRENDS IN MORPHOLOGICAL TECHNIQUES**
**John E. Johnson, Jr., Ph.D.**
Gerontology Research Center
National Institute on Aging
Baltimore City Hospitals
Baltimore, Maryland

**NUTRITIONAL APPROACHES TO AGING RESEARCH**
**Gairdner B. Moment, Ph.D.**
Goucher College, and
Gerontology Research Center
National Institute on Aging
Baltimore, Maryland

**ENDOCRINE AND NEUROENDOCRINE MECHANISMS OF AGING**
**Richard C. Adelman, Ph.D.**
University of Michigan
Ann Arbor, Michigan
**George S. Roth, Ph.D.**
Gerontology Research Center
National Institute on Aging
Baltimore City Hospitals
Baltimore, Maryland

**HANDBOOK OF PHARMACOLOGY OF AGING**
**Paula B. Goldberg, Ph.D. and**
**Jay Robert, Ph.D.**
The Medical College of Pennsylvania
Philadelphia, Pennsylvania

**ALTERED PROTEINS AND AGING**
**Richard C. Adelman, Ph.D.**
University of Michigan
Ann Arbor, Michigan
**George S. Roth, Ph.D.**
Gerontology Research Center
National Institute on Aging
Baltimore City Hospitals
Baltimore, Maryland

**HANDBOOK OF CELL BIOLOGY OF AGING**
**Vincent J. Cristofalo, Ph.D.**
The Wistar Institute and
University of Pennsylvania
Philadelphia, Pennsylvania

Additional topics to be covered in this series include Microbiology of Aging, Evolution and Genetics, Animal Models for Aging Research, and Insect Models.

# Invertebrate Models in Aging Research

Editors

## David H. Mitchell, Ph.D.
President
Sensor Diagnostics, Inc.
Pacific Palisades, California

## Thomas E. Johnson, Ph.D.
Assistant Professor
Genetics and Molecular Biology
Molecular Biology and Biochemistry Department
University of California, Irvine
Irvine, California

**CRC Series on Aging**

Editors-in-Chief

**Richard C. Adelman, Ph.D.**
Director
Institute of Gerontology
Professor of Biological Chemistry
University of Michigan
Ann Arbor, Michigan

**George S. Roth, Ph.D.**
Research Chemist
Gerontology Research Center
National Institute on Aging
Baltimore City Hospitals
Baltimore, Maryland

CRC Press, Inc.
Boca Raton, Florida

**Library of Congress Cataloging in Publication Data**
Main entry under title:                    .

Invertebrate models for aging research.

  (CRC series in aging)
  Bibliography: p.
  Includes index.
  1. Aging. 2. Invertebrates--Physiology.
I. Mitchell, D. H.   II. Johnson, T. E.   III. Series.
QP86.I59   1984      592'.0372      83-24046
ISBN 0-8493-5823-X

Direct all inquiries to CRC Press, Inc., 2000 Corporate Blvd., N.W., Boca Raton, Florida, 33431.

© 1984 by CRC Press Inc.

International Standard Book Number 0-8493-5823-X

Library of Congress Card Number 83-24046
Printed in the United States

# INTRODUCTION

The biological basis of aging has interested scientists and enthralled the layman for decades. Recently, thanks in large part to the influx of new support for basic aging research by the National Institute on Aging, the research literature has been filled with studies on the basic biological processes involved in organismic aging. Often these studies utilize an invertebrate model system.

The short lifespan, ease of culture, and inexpensive maintenance costs make many invertebrate systems especially attractive when compared to vertebrate models. One potential problem with the use of any "model" system be it rodent, primate, or invertebrate is that the molecular basis of aging in the model may be due to a different cause than in humans. Many investigators, feeling that this may be true in aging, have chosen to forego the advantages of working on an invertebrate for the security of feeling that the mouse, rat, or primate is more likely to share common aging mechanisms with humans.

The editors have both chosen the nematode *Caenorhabditis elegans* as a model, thereby supporting the view that basic biological mechanisms will cross species barriers, even between diverse phyla. Dozens, if not hundreds of examples of identical mechanisms could be picked ranging from the universality of the genetic code to the apparent ubiquity of hormonal and phermonal signalling mechanisms. Indeed the overwhelming message is not one of diversity but rather of commonality between different life forms. One instance of an aging phenomenon which is universal wherever the experiments have been performed (i.e., life prolongation resulting from underfeeding or caloric restriction) hints at a universal mechanism underlying this phenomenon.

If there are common molecular processes leading to organismic aging, then it is wise to choose an organism where these processes can be easily studied using many different approaches. Most, if not all, of the organisms described in this book offer multiple avenues of study. The protozoa and the nematode both offer aging models which can be approached at the molecular, cellular, genetic, and organismic levels. Indeed, the nematode is the only species extant where laboratory-derived, long-lived variants are currently being studied.

We have arranged the volume primarily in an order of increasing biological complexity. Thus, Dr. Joan Smith-Sonneborn's chapter on protozoa introduces model systems for research on cellular aging. Dr. Mary Anne Brock discusses the use of Cnidarians in studies of populations as well as individuals. Next are two chapters describing the biology and aging of nematodes by Drs. Michael R. Klass and Thomas E. Johnson; the emphasis of these chapters is on the use of *Caenorhabditis elegans* as a system for the study of the basic biology of metazoan aging and a review of current work on the nematode in other areas relevant to aging research. Drs. William Kristan, David Weisblat, and Tricia Radojcic describe the leech as a model for the aging of the nervous system and Dr. Edwin Cooper describes the earthworm as a model for aging of the immune system. Dr. Michael Kennish presents an overview of the clam, *Mercenaria mercenaria* and the use of the shell in the determination of life-history patterns in wild populations. Finally, Drs. Maynard Makman and George Stefano describe marine mussels and the *Octopus* as models of neuronal aging and touch on an interesting example of programmed senescence in the *Octopus*.

## ACKNOWLEDGMENT

The editors wish to thank Dr. Gale A. Granger, Carl W. Cotman, and Victoria J. Simpson for editorial assistance.

## EDITORS-IN-CHIEF

**Richard C. Adelman, Ph.D.,** is currently Director of the Institute of Gerontology at the University of Michigan, Ann Arbor, as well as Professor of Biological Chemistry in the Medical School. An active gerontologist for more than 10 years, he has achieved international prominence as a researcher, educator, and administrator. These accomplishments span a broad spectrum of activities ranging from the traditional disciplinary interests of the research biologist to the advocacy, implementation, and administration of multidisciplinary issues of public policy of concern to elderly people. He is the author and/or editor of more than 95 publications, including original research papers in refereed journals, review chapters, and books. His research efforts have been supported by grants from the National Institutes of Health for the past 13 consecutive years, and he continues to serve as an invited speaker at seminar programs, symposiums, and workshops all over the world. He is the recipient of the IntraScience Research Foundation Medalist Award, an annual research prize awarded by peer evaluation for major advances in newly emerging areas of the life sciences; and the recipient of an Established Investigatorship of the American Heart Association.

Dr. Adelman serves on the editorial boards of the *Journal of Gerontology, Mechanisms of Ageing and Development,* and *Gerontological Abstracts.* He chaired a subcommittee of the National Academy of Sciences Committee on Animal Models for Aging Research. As an active Fellow of the Gerontological Society, he was Chairman of the Biological Sciences section; a past Chairman of the Society Public Policy Committee; and is currently Chairman of the Research, Education and Practice Committee. He serves on National Advisory Committees which impact on diverse key issues dealing with the elderly, including a 4-year appointment as member of the NIH Study Section on Pathobiological Chemistry; the Executive Committee of the Health Resources Administration Project on publication of the recent edition of *Working with Older People — A Guide to Practice;* and a 4-year appointment on the Veterans Administration Advisory Council for Geriatrics and Gerontology.

**George S. Roth, Ph.D.,** is chief of the Molecular Physiology and Genetics Section of the Gerontology Research Center of the National Institute on Aging in Baltimore, Md., where he has been affiliated since 1972. Dr. Roth received his B.S. in biology from Villanova University in 1968 and his Ph.D. in Microbiology from Temple University School of Medicine in 1971. He received postdoctoral training in Biochemistry at the Fels Research Institute in Philadelphia, Pa. Dr Roth has also been associated with the graduate schools of Georgetown University and George Washington University where he has sponsored two Ph.D. students.

He has published more than 100 papers in the area of aging and hormone/neurotransmitter action, and has lectured, organized meetings, and chaired sessions throughout the world on this subject.

Dr. Roth's other activities include fellowship in the Gerontological Society of America, where he has served in numerous capacities, including chairmanship of the 1979 midyear conference on "Functional Status and Aging". He is a past Chairman of the Biological Sciences Section and a past Vice President of the Society. He has three times been selected as an exchange scientist by the National Academy of Sciences and in this capcity has established liasons with gerontologists, endocrinologists, and biochemists in several Eastern European countries. Dr. Roth serves as an editor of *Neurobiology of Aging, The Journal of Gerontology and Experimental Aging Research,* and is a frequent reviewer for many other journals, including *Mechanisms in Aging and Development, Life Sciences, Sciences,* and *Endocrinology.* He also serves as a reviewer for several funding agencies (including the National Science Foundation) and the research advisory boards for several medical schools. In 1981 Dr. Roth was awarded the Annual Research Award of the American Aging Association. He has also received the NIH Merit Award, and is currently chairman of the Gordon Research Conference on the Biology of Aging.

## THE EDITORS

**Thomas E. Johnson, Ph.D.,** is an Assistant Professor of Genetics in the Department of Molecular Biology and Biochemistry at the University of California at Irvine, California.

Dr. Johnson graduated from the Massachusetts Institute of Technology with a B.Sc. degree in Life Science in 1970. He received a Ph.D. in Genetics from the University of Washington in 1975. Dr. Johnson was a Postdoctoral for 2 years in the Department of Genetics, Development, and Physiology at Cornell University, and a Research Associate in the Department of Molecular, Cellular, and Developmental Biology at the University of Colorado from 1977 to 1982. He became a Fellow of the Insitute for Behavioral Genetics at the University of Colorado in 1981 and an Assistant Professor at the University of California at Irvine in 1982.

Dr. Johnson has published numerous papers in both the genetics of development and the genetics of aging. He has received research support from the National Science Foundation, and the National Institute on Aging, and has served as a reviewer in the fields of genetics, development biology, and gerontology. Dr. Johnson is most recently involved in research on senescence using the nematode, *Caenorhabditis elegans*. The unique biology of the nematode, including a lack of inbreeding depression and the ability to maintain viable cultures indefinitely in liquid nitrogen, make this system unique for studies on aging.

**David H. Mitchell, Ph.D.,** received his B.A. in Biology, with distinction, from Stanford University in 1967. He received an M.A. and Ph.D. in Molecular Biology and Biochemistry from Harvard University in 1975. His work there, in the laboratory of Jonathan Beckwith, involved genetic analysis of control of gene expression.

Between 1975 and 1981 he worked at Boston Biomedical Research Institute as a post-doctoral fellow and later as staff scientist, using genetic techniques to study the processes of aging in the short-lived roundworm *Caenorhabditis elegans*.

In 1981 he founded Sensor Diagnostics, Inc., a biotechnology firm engaged in the development and application of new detection techniques for analysis of blood, urine, and other fluids of medical and biochemical interest. He is currently President of Sensor Diagnostics, Inc.

Dr. Mitchell is interested in the application of genetic techniques to study the processes of aging and in the relationships between development and aging. He also is interested in the interrelationships of science and society and, in particular, the effects and implications of increasing biochemical knowledge for human health and society as a whole.

He presently divides his time between Los Angeles, California, and Cape Cod, Massachusetts.

# CONTRIBUTORS

**Mary Anne Brock**
Research Biologist
Gerontology Research Center
National Institute on Aging
National Institutes of Health
Baltimore, Maryland

**Edwin L. Cooper**
Professor
Department of Anatomy
School of Medicine
University of California
Los Angeles, California

**Thomas E. Johnson**
Assistant Professor of Genetics and
  Molecular Biology
Department of Molecular Biology and
  Biochemistry
University of California
Irvine, California

**Michael J. Kennish**
Senior Environmental Scientist
GPU Nuclear Corporation
Oyster Creek Nuclear Generating Station
Fork River, New Jersey

**Michael R. Klass**
Assistant Professor of Genetics
Biology Department
University of Houston
Houston, Texas

**William B. Kristan, Jr.**
Associate Professor
Department of Biology
University of California, San Diego
La Jolla, California

**Maynard H. Makman**
Professor of Biochemistry and Molecular
  Pharmacology
Department of Biochemistry
Albert Einstein College of Medicine
Bronx, New York

**Tricia Radojcic**
Postgraduate Research Biologist
Department of Biology
University of California, San Diego
La Jolla, California

**Joan Smith-Sonneborn**
Full Professor
Acting Department Head
Zoology and Physiology Department
University of Wyoming
Laramie, Wyoming

**George B. Stefano**
Associate Professor, Chairman
Department of Biological Sciences
State University of New York
Old Westbury, New York

**David A. Weisblat**
Assistant Professor
Zoology Department
University of California
Berkeley, California

TABLE OF CONTENTS

Chapter 1

# PROTOZOAN AGING

## Joan Smith-Sonneborn

# I. PERSPECTIVES

The phylum Protozoa displays a panorama of valuable species for use as model systems of aging. Different protozoa are adapted to such diverse environments as the hostile watery deep, the shallows of quiet ponds, the white water of rapid streams, the soft muddy bottoms of rivers and lakes, the majestic and violent oceans, and in or on bodies of host organisms.

This review will describe the various protozoan model systems, with emphasis on their unique potential to reveal fundamental regulating mechanisms in eukaryotes and the biological principles which seem to drive their maintenance and survival as species. Our premise is that mechanisms responsible for maintenance and/or loss of vitality are shared by protozoa and man. The rationale for this assumption of universality of regulatory mechanisms follows.

It is estimated that 3 billion years elapsed from the origin of life to the existence of diploid organisms, while only 500 million years served to evolve all the multicellular organisms.[1,2] Surely, fundamental regulatory mechanisms were conserved as evolution proceeded, although each cell or organism need not deploy the same mechanism(s) or mesh them in the same patterns to achieve their biological objectives. Regulatory mechanisms involved in cell aging may be more apparent in unicellular than in multicellular organisms, where tissue-tissue interactions could mask the processes.

Among the evolutionarily diverse protozoa, lifespans range from days to years to immortality.[3] Under constant environmental conditions, certain protozoan exhibit an apparent intrinsic programmed senescence. Senescence is characterized by species-specific limited proliferative capacity or viability of representative members of a clone, of a colony, or of an individual nondividing cell. Limited lifespan among protozoa can begin with fertilization, for the clone derived from a single fertilized cell of some ciliates, with differentiation of a terminal cell type in colonial flagellates, or with the emergence of a nondividing adult form which "buds" new progeny.

In some immortal species, heritable finite lifespans can be induced by altered diet, as in *Amoeba*[4,5] or by inbreeding an outbreeder, as in the ciliate *Tetrahymena*.[6]

The ability of certain cell types to exhibit an apparently endless capability for proliferation is seen in the reproductive cell of colonial flagellates, and in the outbreeding *Tetrahymena*.

Aging in protozoa has been reviewed recently,[7,8] and only highlights of aspects of the model systems will be presented here for clonal aging, colonial aging, individual aging, and environmentally induced alteration in lifespan.

# II. CLONAL AGING IN CILIATES

## A. Biological Background

A clone is a multicellular unit derived from a single cell. The clonal unit is composed of cells separated in space. Clonal aging refers to the decreased probability that a cell will give rise to a viable product at the next cell division as time (in cell divisions or days [age]) evolved since the last fertilization.[9,10] Fertilization can occur by conjugation (mating with an opposite mating type), autogamy (self-fertilization), or selfing (mating with a clone). Members of a clone pass synchronously and predictably through developmental periods of immaturity (when cells cannot mate), maturity (when cells can mate), senescence (when the probability that a viable cell will be produced at the next cell division is reduced), and finally death. The different developmental periods are found at a species–specific number of cell divisions postfertilization. A specific cytoplasmic protein, immaturin, represses sexual activity in *Paramecium* and may regulate expression of maturity in these cells.[11,12]

A prerequisite for clonal aging studies is thorough investigation of the life cycle of the organism under consideration. The "Paramecium Methuselah" strain, once thought to be immortal,[13] was discovered to be a succession of clones undergoing autogamy or self-fertilization.[10] In the absence of fertilization, the clone undergoes progressive deterioration, and death of all members of the clone follows.

In general, the ciliates contain two kinds of nuclei, the micronucleus or germ line and the macronucleus or somatic line. The micronuclei represent the sexual repository of genetic information for the next generation and show little transcriptional activity during the asexual (binary cell division) cycle,[10,14,15-19] and function during the fertilization process. The macronucleus dictates the metabolic activity of the cell.[19,20,22] The transcriptionally active macronucleus of *Tetrahymena* has been shown to have acetylated histone, whereas the transcriptionally inactive micronucleus does not.[23] The micronuclei form gametes which fuse to form the zygote for differentiation of new nuclei in the typical ciliate fertilization processes. The protozoans offer a valuable model system of inactive and active genetic information in the different nuclei in a common cytoplasm. The differentiation of micro- and macronuclei occurs during fertilization.

During conjugation in the ciliate *Paramecium*, for example, cells come together, side by side, and eventually exchange nuclear gametes through a temporary transfer organelle. Comparable to nuclei of sex cells of higher organisms, the micronuclei undergo meiosis to generate gametes which function during fertilization. In each mate, meiosis of the two micronuclei produces eight gametes, seven of which disintegrate, and only one haploid gamete is retained. This gamete duplicates into two identical haploid gametes. One remains stationary; the other migrates into the mate and fuses with the mate's stationary gamete. Thus, in true conjugation, the two members of a mating pair become genetically identical, each with half of its nuclei from its own duplicated haploid complement and half from its partners. The details of the developmental events postfertilization differ with species, but normally the zygote undergoes differentiation of a new micro- and macronucleus for progeny cells. In autogamy, fertilization occurs in a single cell. After meiosis, one haploid gamete is retained and duplicated into two identical gametes which fuse to form a homozygous diploid gamete (for review of ciliate life cycles and genetics, see Preer.[24]) If conjugation is in progress, mutual exchange of gametes occurs; in autogamy, the identical gametes fuse to form the zygote nucleus (the synkaryon). This synkaryon divides twice mitotically to form four genotypically identical diploid nuclei. In *Tetrahymena*, one disintegrates, one stays a micronucleus, and the remaining two develop into macronuclei (for review see Sonneborn[25]). The position of the synkaryonic products within the cell can determine which nuclear differentiation is favored,[21,26] and recent studies indicate that removal of the anterior anlage generally results in cells without micronuclei.[27] The posterior anlage undergo macronuclear development, and several rounds of DNA replication occur within the first two cell divisions to provide a *Paramecium* macronucleus with 800 times the DNA content of the haploid micronucleus. The ancestral macronucleus can function during the development of the new filial macronucleus.[28-30] The RNA produced by the ancestral macronucleus represents a maternal cytoplasmic environment reminiscent of the stored RNA in oocyte cytoplasm in higher organisms. If the new macronucleus fails to form, the ancestral macronucleus can persist (*Tetrahymena*), or fragments of the ancestral macronucleus can regenerate (*Paramecium*), termed macronuclear retention and regeneration in the respective genus.

The genetic consequences of the fertilization process in *Paramecium* are as follows: if the genotypes of the two gametes are denoted "A" and "A" in one mate and "a" and "a" in its partner, then, after mutual exchange of gametes, both members of the pair become Aa. In autogamy, the two identical gametes fuse in a single cell yielding either AA or aa from an Aa parent cell; half AA and half aa progeny result from a pool of hybrid parents. If a young and an old cell mate, genetically identical pairs can result in their respective young or old cytoplasm.

It is also possible to mate genetically identical cells expressing different mating or antigenic types, i.e., differentiated cells, and to influence the determination of mating or antigenic

## Table 1
## LIFESPAN OF VARIOUS PROTOZOA

Maximal Lifespan

| Organism | Cell divisions | Approximate months | Ref. |
|---|---|---|---|
| *Paramecium Aurelia* **Complex** | | | |
| P. primaurelia | 340 | 4 | 10, 108 |
| P. biaurelia | 286 | 5 | 10, 108 |
| P. tetraurelia | 244 | 2 | 10, 36, 108 |
| P. caudatum | 600—700 | 7 | 109 |
| P. bursaria | — | 80 | 110 |
| P. multimicronucleatum | — | 180 | 108 |
| Spathidium muscicola | 134 | — | 111 |
| S. spathula | 658 | — | 111 |
| Tokophrya lemnarum | 800 | — | 66 |
| **Euplotes** | | | |
| E. woodruffi (fresh water) | 208 | — | 57, 112, 113 |
| E. woodruffi (marine) | 339 | 5 | 57, 112, 113 |
| E. crassus | 1000 | — | 114 |
| E. patella | 1300 | 50 | 58, 115 |
| Oxytricha bifari | — | 30 | 116 |
| Stylonychia pustula | 316 | 4 | 9 |
| Tetrahymena (after inbreeding) | 40—1500 | — | 72-74 |

type by environmental conditions during the critical period of development of the new macronucleus.[25]

Methodologies used to study clonal aging vary with investigator, but in general, cells after fertilization are placed in a growth medium and transferred daily or at regular intervals to fresh medium. The total number of cell divisions (or cycles of bud-parent-bud in *Tokophrya)* from the origin of the clone at fertilization to death represents the lifespan. Daily isolations of several sublines (representatives) of a clone allow observation of the number of cells produced by each single cell ($\log_2$ of that number is the daily fission rate) each day after fertilization. A single cell from a population of cells produced in the previous 24 hr in a subline is reisolated on successive days, and the sum of the fissions per day is the age of the cell. When an isolated cell fails to survive on any day, the sum of the number of cell divisions or days since fertilization represents the lifespan of that subline. The mean lifespan is the average clonal age for all sublines in a control or experimental group. Maximal lifespan is the highest number of cell divisions or days attained by any subline within an experimental or control group. Clonal aging has been documented in certain species of *Paramecium*, *Euplotes, Stylonychia, Tokophrya,* and *Spathidium* and can be induced either by inbreeding or in aberrant lines of *Tetrahymena* (Table 1). See review, Smith-Sonneborn.[8,31]

## B. Paramecium

In the *Paramecium aurelia* species complex, the lifespan of different species varies from 200 to 350 cell divisions (Table 1). Progeny survival after autogamy declines with increased parental age as does the lifespan of the resultant clone.[10,32-36] Age-related changes are outlined below and in Table 2.

Morphological changes which occur include variable number of micronuclei,[37,38] chromosomal aberrations,[39] abnormal macronuclei,[40] increased abnormal mitochondria,[41] the

## Table 2
## AGE-RELATED CHANGES IN CILIATES

|  | Ref. |
|---|---|
| Decreases in: |  |
|     DNA synthesis in *P. aurelia* | 55 |
|     RNA synthesis in *P. aurelia* | 51 |
|     Food vacuole Number in *P. aurelia* | 48 |
|     Macronuclear DNA in *P. aurelia, P. caudatum,* and *P. bursaria* | 49-52 |
|     Progeny viability in *P. aurelia, Euplotes patella,* and *P. caudatum* | 10, 32-36, 44, 47, 57, 58 |
|     Nucleolar volumes in *P. aurelia* | 123 |
|     DNA amount in *P. aurelia* | 49-52 |
|     Ribosomal clusters in *P. aurelia* | 124 |
|     Food vacuole number and synthesis in *P. aurelia* | 48 |
|     Fission rate in *P. aurelia, P. caudatum, Euplotes patella,* and *E. woodruffi* (marine) | 10, 36, 58, 109, 113 |
|     DNA template activity in *P. aurelia* | 55 |
|     RNA synthesis in *P. aurelia* | 51 |
| Increases in: |  |
|     Variable number of micronuclei in *P. aurelia* | 37, 38 |
|     Chromosomal aberrations in *P. aurelia* | 10, 35, 38, 39, 49, 119, 120 |
|     Fused macronuclear chromatin in *P. aurelia* | 40 |
|     Abnormal mitochondria in *P. aurelia* and *Tokophrya infusionum* | 41, 69 |
|     Age pigments in *P. aurelia* | 41 |
|     Nucleolar volume in *P. aurelia* | 42 |
|     Cytoplasmic toxicity in *P. aurelia, P. bursaria,* and *E. patella* | 35, 44, 110, 115, 121 |
|     UV sensitivity in *P. aurelia* | 7, 45 |
|     Lysosomal activity in *P. aurelia* | 40, 41 |
|     Micronuclear aberrations in *P. aurelia* and *E. patella* | 21, 38, 39, 43, 58, 119-121 |
|     Variability in number of micronuclei in *P. aurelia* | 37, 38 |
|     Invagination in macronucleus in *P. aurelia* | 40 |
|     Abnormal macronuclear appearance in *P. aurelia* | 40, 49, 52, 122 |
|     Toxicity of cytoplasm in *P. aurelia* and *Euplotes patella* | 35, 44, 47, 58 |
|     Lysosomal activity in *P. aurelia* and *Tokophrya infusionum* | 40, 41, 68 |
|     Sensitivity to ultraviolet in *P. aurelia* | 7, 45 |
|     Sensitivity to X-rays in *P. aurelia* | 101, 125 |
| No change: |  |
|     Copper sensitivity in *P. aurelia* | 43 |
|     DNA polymerase activity in *P. aurelia* | 44 |
|     Copper tolerance in *P. aurelia* | 43 |

appearance of age pigments,[41] and increase in nucleolar volume.[42] Copper sensitivity and DNA polymerase activity apparently do not change with age.[43,44] There is an increase in ultraviolet (UV) sensitivity[7,45] and lysosomal activity.[40,41] The cytoplasm of old cells acquires toxic characteristics. In old-young crosses, some lethality results from age-correlated micronuclear damage[35] although old-young cytoplasmic incompatibility is also indicated by the observation that abnormalities in the young mate are induced prior to nuclear transfer.[46]

Evidence for the toxic nature of aged cytoplasm was also seen in "merogones" (old cells whose nuclei fail to function but which receive the young nuclear complement). Usually, merogones do not survive, indicating the toxic nature of aged cytoplasm. The occurrence of death without phenotypic lag[37] in *P. primaurelia* implies dominant lesions or cytoplasmic toxicity. Evidence for the ability of a young nucleus to rejuvenate an old cytoplasm was provided by merogones; occasionally, if the young nucleus survived the crisis, vigorous progeny could be obtained.[35] Evidence for the ability of a young cytoplasm to rejuvenate an old nucleus was provided in *P. caudatum*. (Micronuclei transferred from aged cells to young cells produced more viable offspring than when the aged nuclei developed in the aged cell). In very old cells, a point of "no return" was found (when transfer of the old nucleus to a young cytoplasm had no beneficial effect).[47] It is possible that (1) the old nucleus which could be rejuvenated in young cytoplasm was not damaged but became injured only when development occurred in aged cytoplasm; or (2) the old micronucleus was damaged and could be repaired in young cytoplasm. In any case, in order to function to produce many vigorous offspring, the old nuclei needed the young cytoplasm.

In the clonal-aging studies of *Paramecium*, the following age-related changes have been described (Table 2). There is a decline in cell division rate[10] and in endocytic capacity.[48] The amount of DNA in the macronucleus was found to decline in the first half of the life cycle in *P. tetraurelia*[49-51] and in *P. bursaria*.[50] *P. caudatum* does show a decline in macronuclear DNA in the last third of the life cycle. Macronuclear DNA drops to 1/5 of the initial DNA content at the onset of senescent traits.[52] Certain *Tetrahymena* also show a decrease in macronuclear DNA.[53] DNA and RNA synthesis rates decline with age[51,55] as does progeny viability.

## C. *Euplotes* and Other Hypotrichs

In *Euplotes crassa*, age-related loss of phenotypic dominance was observed.[56] Using *E. woodruffi* in old-young crosses,[57] both members of the mating pair showed abnormalities, indicating that the aged phenotype was imposed on the young mate. In *E. patella*, abnormalities occurred prior to nuclear exchange, indicating the induction of abnormalities by the cytoplasm.[58] The old mate was neither able to undergo nuclear development nor to form a normal mouth.

Recent studies in *E. aediculatus* indicate that normal macronuclear development is essential for normal lifespan. Two distinct developmental "hurdles" must be passed, one early (a few hours after separation of pairs) and one late (34 to 37 hr after pair separation). The late crisis occurs when the peak number of chromosomes have been replicated. After this time, these ciliates degenerate certain chromosomes. It has been suggested that certain chromosomes function only this once in a lifetime and then are eliminated.[59] The genes which must function in these ciliates are related to cortical or surface morphogenesis of mouth parts. Clearly, if cytoplasmic conditions in an old cell do not favor expression of genes at this critical time, normal function is not possible.

There is general agreement that most DNA sequences in the micronuclear zygote are lost during differentiation to the final somatic macronucleus in *Euplotes* and other hypotrichs like *Stylonychia mytilus* and *Oxytricha*.[60,61] Both *Paramecium* and *Tetrahymena*, however, seem to keep most of the micronuclear sequences in the macronucleus.[28,54,62-64] *Stylonychia* and *Oxytricha* may have developmental hurdles like those of *Euplotes*. The recent studies also provide a rationale for the micronucleus carrying more genes than are incorporated into the macronucleus, i.e., certain genes only need to function early in development. Once this function is performed, these genes do not need to be carried in the somatic macronucleus and can be preserved for the next generation in the relatively transcriptionally inert micronucleus.

## D. *Tokophrya*

Clonal aging in *Tokophrya lemnarum* was noted at a given interval after fertilization of parent cells. When older cells mated, abnormal pairs were noted. Immature, mature, and senescent cells were also found at intervals postconjugation.[65,66] In addition to clonal aging, the striking morphological differences in the possibly stalked parent cell and the swimming bud it produces allow study of the lifetime of the individual adult cell.[67-71]

## E. *Tetrahymena*

When the long-lived outbreeding strains of *Tetrahymena* were inbred, progeny resulted which either (1) died within the first 10 fissions after conjugation; (2) retained the old macronucleus; or (3) showed early onset of micronuclear abnormalities within 500 to 1500 fissions.[72,73] With stringent selection of viable progeny, viable inbred lines were obtained.

Naturally occurring, apparently immortal, micronucleated and amicronucleated strains of *Tetrahymena* exist; conversion from micro- to amicronucleated strains does not result in viable cells.[74] Examination of 8 different strains of *Tetrahymena* for 800 generations revealed that all but one were nonaging lines.[75] When deterioration arises in strains of *Tetrahymena*, it is usually in the "silent" micronucleus. The occurrence of germ-line erosion appears to be random and not a programmed event.[76]

*Tetrahymena* offers the advantage of a species with only five chromosomes. The genetics of *Tetrahymena* advanced explosively when the silence of the germ-line nucleus was exploited to produce strains missing both copies of one of the five chromosomes.[77]

## III. COLONIAL AGING

### A. Definition

Colonial green flagellates are algae by virtue of their pigments, and protozoans, because the architecture of the colony is composed of unicellular, flagellated (long whip-like organelles of locomotion) building blocks (for review, see Coleman[78] and Weise[79]). These protozoans exhibit embryogenesis, including inversion stages with separation of somatic (finite lifespan) and germ-line (infinite lifespan) cells.

### B. *Volvox*

Colonial aging is detected as a loss of function of individual somatic cell types within a colony. In certain *Volvox* colonies, more than 99% of the cells are somatic and undergo synchronous programmed senescence and cell death every generation.[80] A small number of reproductive cells survive to produce the next generation. The potential reproductive cells are set aside at a particular stage by unequal cell division.[81,82] The somatic and reproductive cells exhibit different electrophoretic patterns of polypeptides and the patterns change when senescent characteristics are first evident in somatic cells.[80]

Electron microscopic studies of age-dependent changes in the terminally differentiated somatic cells of the colonial green algae *V. carteri* revealed disorganization of chloroplast structure, decline in cytoplasmic ribosomes, and accumulation of lipid bodies in the cytoplasm of the older cells. Since these changes are typical of those noted in starved cells, the possibility that an inability to take up or utilize nutrients may cause or contribute to senescence in terminally differentiated somatic cells is suggested.[83]

Colonies can be propagated indefinitely by vegetative reproduction in haploid colonial green flagellates. The only diploid stage in the life cycle is the zygote which, like the spores of bacteria, can remain dormant and genetically preserved for years.

The colonial green flagellates provide an elegant system for regulation of cell proliferation potential. *Volvox* mutants which fail to segregate replicative and somatic forms are available. This array of mutants suggests that proliferative capacity involves several regulatory genes.[84]

## IV. DIET-INDUCED ALTERATION OF LIFESPAN

### A. *Tokophrya*

The adult *Tokophrya* has been used to assay effects of feeding on longevity. The adult lives 10 days in *T. infusionum*[69] and 16 days in *T. lemnarum*.[66] After formation of the adult, there is a short immaturity period when buds are not produced, a mature period of bud production, and a senescent period when no buds can be formed. *Tokophrya* feeds on living prey by sucking host contents through long slender tubes (tentacles). When given excess food, a *Tokophrya* can feed simultaneously with all of its 50 tentacles. The young gluttonous

*Tokophrya* at the age of 4 days resembles a normal organism twice that age, and can assume giant proportions (120 times normal). The adult *Tokophrya* does not have an organelle for excretion of solid wastes; the opening to the outside is for the release of buds.[71] Therefore, the adult *Tokophrya* can serve as a model system for effects of accumulated digestive wastes on nondividing cell survival. As might be predicted in these adults with defective excretory capacity, over-feeding shortens life, intermittent starvation prolongs life, and starvation shortens life.[67-69]

### B. *Amoeba*

The apparent immortality of *Amoeba*, seen when these cells are grown exponentially, is a precarious state. Growth of *A. proteus* or *A. discoides* in a maintenance diet induces a finite lifespan, and the "lifespanning or aging" phenomenon in the free-living protozoa *A. proteus* was reported.[4,5] When cells were exposed for 2 to 9 weeks to a maintenance diet, a defined lifespan which varied from 30 days to 30 weeks ensued. Cells with defined lifespan show two major forms of behavior. Type A stem-line growth was characterized by production of one viable and one nonviable product (one cell which died within a day or two after cell division) until eventually both daughter cells died. Type B cells produced viable daughters until the time of clonal death. Microsurgical transfers revealed that both nucleus and cytoplasm carried "spanning" information.[85-87]

## V. STARVATION IN PROTOZOA

In contrast to the intrinsic clonal aging seen in ciliates, morphological and physiological changes, followed by cell death, can be induced by starvation or "cultural aging". This is distinctly different from naturally occurring aging in a constant environment. Abnormalities are inflicted on cells due to unfavorable environmental conditions rather than sequentially expressed under constant external environmental conditions as in clonal aging.[10] When starved, dinoflagellates have been found with fused chromatin in the nucleus;[88] *Euglena* cytoplasm contained heavily pigmented bodies and membrane fragments;[89] *Tetrahymena* cytoplasm was vacuolized and showed increased numbers of mitochondria and lipid droplets;[90] *Ochromonas* cytoplasm was characterized by increased lysosome activity, increased lipid vacuoles, and disorganization.[91] In *Paramecium caudatum*, a decrease in digestive vacuoles was concomitant with an increase of lysosomes, autophagosomes, and lipofuscin-like or aging-pigment-like granules.[92]

Variations in temperature, oxygen, tension, and the culture age used to inoculate fresh axenic medium all influenced biochemical, physiological, and ultrastructural relationships.[93] The ability of some small proportions of cells in early death phase to exclude accumulated wastes and return to a healthy state when transferred to fresh medium is of interest both to understanding regulation of phagosome-lysosome systems of cells[93] and the recovery potential from starvation as a function of clonal age.

Some morphological changes seen in exogenously starved cells mimicked the structural changes seen in aged *Paramecium, Euplotes,* and *Tokophrya* adults. The morphological changes in the macronuclei of aged paramecia[40] have also been noted in young starved cells.[42] Aged paramecia, even in the presence of excess food, can show few or no food vacuoles.[48] Starvation may not be a cause of aging, but it appears to be a proximal cause of death. Autolytic processes, stimulated at the termination of cell life, mimic those of young cells when starved.

Studies on the effects of starvation provide models to explore reversibility of the effects of starvation and perhaps provide a means to delay terminal aging stages.

# VI. ENVIRONMENTAL MODULATION OF LONGEVITY

The use of environment to alter gene activity is a potential strategy to vary lifespan. A rational basis for the feasibility of this approach stems from evidence that intracellular environment and extracellular conditions can trigger transitory or even permanent changes in expression of one of an alternative set of genes. Therefore, the expression of a particular gene(s) can be influenced by the environment.[25,94] It is hoped that levels of gene expression may also be subject to environmental modulation.

The ciliates represent a genetic system in which genotypically identical cells exhibit cellular differences based on stable nuclear differentiations.[25] This is not totally unlike genotypically identical cells of multicellular organisms expressing different genes for cell function in different tissues. The stable nuclear differentiations can be induced during the sensitive period when new zygotic macronuclei are developing and can be affected by alterations in temperature, nutrition, ion concentration, and cytoplasmic factors.[25,94-97] Shifts in gene expression can also occur later in the life cycle. However, the potential for shifts may only be a restricted part of the range potential of an undifferentiated nucleus.[98] A change in the environment during the vegetative life cycle can induce expression of one of a range of antigenic types. The specific new type that appears depends on both the old and new environment and the previous type expressed.[99,100] That environment can influence gene expression provides a rationale for environmentally influenced genes which effect longevity.

With respect to environmentally induced alteration of longevity, radiation effects have been monitored in *Paramecium*. Lifespan can be shortened by repeated exposure to massve doses of X-irradiation,[101] UV-irradiation,[7] and $^{60}$Co radiation.[102] Since sensitivity to UV-irradiation increases with age,[45] there may be an age-related loss of excision repair capacity.

The photoreactivation repair of UV-induced damage can result in a significant increase in lifespan. It is assumed that the UV damage stimulates DNA repair and when photoreactivation repairs UV damage, the cell possesses excess repair capacity which can correct some age damage. Cells which survived the cytotoxic effects of $^{60}$Co irradiation also lived longer than control cells,[102] which may be in part due to some induced repair capability.

An alternative approach for increasing repair capacity is the detection of drugs which stimulate repair. Drugs which enhance repair may be expected to protect cells against UV mutagenesis. Conversely, inhibitors of repair should increase expression of UV mutagenesis. The bioassay is thus used to detect drugs which modulate repair. The antibiotic novobiocin, is one drug found to increase sensitivity of *Paramecium* to UV irradiation.[105] Novobiocin inhibits endonuclease recognition of UV lesions[103] and DNA polymerase.[104] The ability of photoreactivation to reverse the novobiocin-enhanced mutagenesis[106,107] indirectly implies that the biological effect is due to inhibition of DNA repair. Novobiocin only, at 1 μg/mℓ, shortens lifespan. A mitosis-stimulating drug in plants, kinetin, was found to decrease the effects of UV mutagenesis in *Paramecium*. Kinetin also decreases the novobiocin-enhanced UV damage.[107] The inhibitor of UV repair, novobiocin, shortens lifespan in *Paramecium* whereas the enhancer of repair, kinetin, was found to increase lifespan in the later half of the lifespan. The concordance of inhibitor-of-repair with shortened lifespan and enhancer-of-repair with increased lifespan is evidence for the importance of DNA integrity in lifespan determination.

# VII. SUMMARY

The protozoa offer several elegant models of programmed cellular senescence with species-specific limited proliferation potential. Parental cell parts or types are conserved and are passed to the successive generation. The micronuclear gametes in ciliates differentiate new nuclei and reproductive cells generate somatic cells in the colonial flagellates.

Immortality can be a precarious state in the apparently infinite lifespan of *Tetrahymena* and *Amoeba*. Inbreeding the outbreeder can induce limited lifespan in *Tetrahymena* and exposure to a maintenance diet can induce limited lifespan in *Amoeba*.

Each species may employ mechanisms integrated in a particular way to ensure survival. The array of regulatory mechanisms need not be large, since meshing a few mechanisms in a variety of sequences can generate a multitude of unique programs. All cells may share mechanisms sensitive to environmental perturbations. Cellular reserves may be available by environmental stimulation for correction or retardation of age damage and may lead to the ability to minimize pathologies associated with senescence.

# REFERENCES

1. **Margulis, L.,** *Origin of Eukaryotic Cells,* Yale University Press, New Haven, 1970.
2. **Ohno, S.,** *Evolution of Gene Duplication,* Springer-Verlag, Basel, 1970.
3. **Sonneborn, T. M.,** Enormous differences in length of life of closely related ciliates and their significance, in *The Biology of Aging,* Strehler, B. L., Ed., Am. Inst. Biol. Symp., Waverly Press, Baltimore, 1960, 289.
4. **Muggleton, A. and Danielli, J. F.,** Aging of *Amoeba proteus* and *A. discoides* cells, *Nature (London),* 181, 1783, 1958.
5. **Muggleton, A. and Danielli, J. F.,** Inheritance of the "life-spanning" phenomenon in *Amoeba proteus, Exp. Cell Res.,* 49, 116, 1968.
6. **Nanney, D. L.,** Ciliate genetics: patterns and programs of gene action, *Ann. Rev. Genet.,* 2, 121, 1968.
7. **Smith-Sonneborn, J.,** DNA repair and longevity assurance in *Paramecium tetraurelia, Science,* 203, 1115, 1979.
8. **Smith-Sonneborn, J.,** Genetics and aging in protozoa, *Int. Rev. Cytol.,* 73, 319, 1981.
9. **Jennings, H. S.,** Genetics of the protozoa, *Biblio. Genet.,* 5, 105, 1929.
10. **Sonneborn, T. M.,** The relation of autogamy to senescence and rejuvenescence in *Paramecium aurelia, J. Protozool.,* 1, 38, 1954.
11. **Haga, N. and Hiwatashi, K.,** A protein called immaturin controlling sexual immaturity in *Paramecium, Nature (London),* 289, 177, 1981.
12. **Miwa, I., Haga, N., and Hiwatashi, K.,** Immaturity substances: material basis for immaturity in *Paramecium, J. Cell Sci.,* 19, 360, 1975.
13. **Woodruff, L. L.,** *Paramecium aurelia* in pedigree culture for twenty-five years, *Trans. Am. Microsc. Soc.,* 51, 196, 1932.
14. **Sonneborn, T. M.,** Inert nuclei: inactivity of micronulear genes in variety 4 of *Paramecium aurelia, Genetics,* 31 (Abstr.), 231, 1946.
15. **Nobili, R.,** I. Dimorfismo Nucleare die Ciliate: inerzia Vegetativa del Micronucleo, *Acad. Nazi Lincei Rend. Class Sci. Fisiche Mat. Nat. Ser.,* 8(32), 395, 1962.
16. **Pasternak, J.,** Differential genetic activity in *Paramecium aurelia, J. Exp. Zool.,* 165, 395, 1967.
17. **Gorovsky, M. A. and Woodard, J.,** Studies in the nuclear structure and function in *Tetrahymena pyriformis.* I. RNA synthesis in macro and micronuclei, *J. Cell Biol.,* 42, 673, 1969.
18. **Murti, K. G. and Prescott, D. M.,** Macronuclear ribonucleic acid in *Tetrahymena pyriformis, J. Cell Biol.,* 47, 460, 1970.
19. **Klass, M.,** DNA Template Activity of Ethanol-Fixed Micronulcei of *Paramecium aurelia,* Ph.D. dissertation, University of Wyoming, Laramie, 1974.
20. **Sonneborn, T. M.,** Recent advances in the genetics of *Paramecium* and *Euplotes, Adv. Genet.,* 1, 263, 1947.
21. **Sonneborn, T. M.,** Patterns of nucleocytoplasmic interactions in *Paramecium, Caryologia,* 6, 307, 1954a.
22. **Siegel, R. W.,** New results on the genetics of mating types in *Paramecium bursaria, Genet. Res.,* 4, 132, 1963.
23. **Gorovsky, M. A., Pleger, G. L., Kevert, J. B., and Johmann, C. A.,** Studies on histone fraction F2a1 in macro and micronuclei of *Tetrahymena pyriformis, J. Cell Biol.,* 57, 773, 1973.
24. **Preer, J. R., Jr.,** Genetics of the protozoa, in *Research in Protozoology,* Vol. 3, Chen, T. T., Ed., Pergamon Press, N.Y., 1968, 139.
25. **Sonneborn, T. M.,** Genetics of cellular differentiation: stable nuclear differentiation in eucaryote unicells, *Annu. Rev. Genet.,* 11, 349, 1977.

26. **Nanney, D. L.,** Nucleo-cytoplasmic interaction during conjugation in *Tetrahymena, Biol. Bull.,* 105, 133, 1953.
27. **Mikami, K.,** Nuclear transplant studies on the reduction in numbers of presumptive germ nuclei in exconjugants of *Paramecium caudatum, J. Cell Sci.,* 56, 453, 1982.
28. **Berger, J. D.,** Nuclear differentiation and nucleic acid synthesis in well-fed exconjugants of *Paramecium aurelia, Chromosoma,* 42, 247, 1973.
29. **Berger, J. D.,** Selective autolysis of nuclei as a source of DNA precursors in *Paramecium aurelia* exconjugants, *J. Protozool.,* 21, 124, 1974.
30. **Berger, J. D.,** Gene expression and phenotype change in *Paramecium* exconjugants, *Genet. Res.,* 27, 123, 1976.
31. **Smith-Sonneborn, J.,** Aging in protozoa, in *Review of Biological Research in Aging,* Vol. 1, Rothstein, M., Ed., Alan Liss, N.Y., 1983, 29.
32. **Pierson, B. F.,** The relation of mortality after endomixis to the prior endomicitic interval in *Paramecium aurelia, Biol. Bull.,* 74, 235, 1938.
33. **Sonneborn, T. M. and Schneller, M. V.,** Are there cumulative effects of parental age transmissible through sexual reproduction in variety 4 of *Paramecium aurelia?, J. Protozool.,* 2(Suppl.), 6, 1955.
34. **Sonneborn, T. M. and Schneller, M. V.,** The basis of aging in variety 4 of *Paramecium aurelia, J. Protozool.,* 2, 6, 1955a.
35. **Sonneborn, T. M. and Schneller, M. V.,** Measures of the rate and amount of aging on the unicellular level, in *The Biology of Aging,* Strehler, B. L., Ed., Am. Inst. Biol. Sci. Symp. 6, Waverly Press, Baltimore, 1960, 290.
36. **Smith-Sonneborn, J., Klass, M., and Cotton, D.,** Parental age and lifespan versus progeny lifespan in *Paramecium, J. Cell Sci.,* 14, 691, 1974.
37. **Mitchison, M. A.,** Evidence against micronuclear mutations as the sole basis for death at fertilization in aged, and in the progeny of ultraviolet irradiated *Paramecium aurelia, Genetics,* 40, 61, 1955.
38. **Sonneborn, T. M. and Dippell, R. V.,** Cellular changes with age in *Paramecium,* in *The Biology of Aging,* Strehler, B. L., Ed., Am. Inst. Biol. Sci. Symp. 6, Waverly Press, N.Y., 1960, 285.
39. **Dippell, R. V.,** Some cytological aspects of aging in a variety of *Paramecium aurelia, J. Protozool.,* 2(Suppl.), 7, 1955.
40. **Sundararaman, V. and Cummings, D. J.,** Morphological changes in aging cell lines of *Paramecium aurelia.* II. Macronuclear alterations, *Mech. Ageing Dev.,* 5, 325, 1976.
41. **Sundararaman, V. and Cummings, D. J.,** Morphological changes inn aging cell lines of *Paramecium aurelia.* I. Alterations in the cytoplasm, *Mech. Ageing Dev.,* 5, 139, 1976a.
42. **Heifetz, S. and Smith-Sonneborn, J.,** Nucleolar changes in aging and autogamous *Paramecium tetraurelia, Mech. Age. Dev.,* 16, 255, 1981.
43. **Nyberg, D.,** Copper tolerance and segregation distortion in aged *Paramecium, Exp. Gerontol.,* 13, 431, 1978.
44. **Williams, T. J. and Smith-Sonneborn, J.,** Induction of DNA polymerase activity by ultraviolet irradiation of *Paramecium tetraurelia, Exp. Gerontol.,* 15, 353, 1980.
45. **Smith-Sonneborn, J.,** Age correlated sensitivity to ultraviolet radiation in *Paramecium, Radiat. Res.,* 46, 64, 1971.
46. **Chen, T. T.,** Conjugation in *Paramecium bursaria.* II. Nuclear phenomena in lethal conjugation between varieties, *J. Morphol.,* 79, 125, 1946.
47. **Karino, S. and Hiwatashi, K.,** Analysis of germinal aging in *Paramecium caudatum* by micronuclear transplantation, *Exp. Cell Res.,* 136, 407, 1981.
48. **Smith-Sonneborn, J. and Rodermel, S. R.,** Loss of endocytic capacity in aging *Paramecium:* the importance of cytoplasmic organelles, *J. Cell Biol.,* 71, 575, 1976.
49. **Schwartz, V. and Meister, H.,** Eine Alterveranderung des Makronucleus von *Paramecium, Z. Naturforsch. Sect. B.,* 28c, 232, 1973.
50. **Schwartz, V. and Meister, H.,** Aging in *Paramecium:* several quantitative aspects, *Arch. Protistenk. Biol.,* 117, 85, 1975.
51. **Klass, M. and Smith-Sonneborn, J.,** Studies in DNA content, RNA synthesis, and DNA template activity in aging cells of *Paramecium aurelia, Exp. Cell Res.,* 98, 63, 1976.
52. **Takagi, Y. and Kanazawa, N.,** Age-associated change in macronuclear DNA content in *Paramecium caudatum, J. Cell Sci.,* 54, 137, 1982.
53. **Doerder, F. P. and Debault, L. E.,** Life cycle variation and regulation of macronuclear DNA content in *Tetrahymena thermophila, Chromosoma,* 69, 1, 1978.
54. **Doerder, F. P. and Debault, L. E.,** Cytofluorimetric analysis of DNA during meiosis, fertilization, and macronuclear development in the ciliate *Tetrahymena pyriformis,* syngen 1, *J. Cell Sci.,* 17, 471, 1975.
55. **Smith-Sonneborn, J. and Klass, M.,** Changes in the DNA synthesis pattern of *Paramecium* with increased clonal age and interfission time, *J. Cell Biol.,* 61, 591, 1974.

56. **Heckmann, K.**, Age-dependent intraclonal conjugation in *Euplotes crassus, J. Exp. Zool.*, 2(165), 269, 1967.
57. **Kosaka, T.**, Age-dependent monsters of macronuclear abnormalities, the length of life, and a change in the fission rate with clonal aging in marine *Euplotes woodruffi, J. Sci. Hiroshima Univ., Ser. B. Div. 1,* 25, 173, 1974.
58. **Katashima, R.**, Several features of aged cells in *Euplotes patella,* syngen 1, *J. Sci. Hiroshima Univ., Ser. B. Div. 1,* 23, 59, 1971.
59. **Kloetzel, J. A.**, Nuclear roles in postconjugant development of the ciliated protozoan *Euplotes aldiculatus, Dev. Biol.*, 83, 20, 1981.
60. **Ammermann, D., Steinbruck, G., van Berger, L., and Hennig, W.**, The development of the macronucleus in the ciliated protozoan *Stylonychia mytilus, Chromosoma,* 45, 401, 1974.
61. **Lauth, M. R., Spear, B. B., Heumann, J. M., and Prescott, D. M.**, DNA of ciliated protozoa: DNA sequence diminution during macronuclear development of *Oxytricha, Cell,* 7, 67, 1976.
62. **Yao, M. C. and Gorovsky, M. A.**, Comparison of the sequence of macro and micronuclei on DNA of *Tetrahymena pyriformis, Chromosoma,* 48, 1, 1974.
63. **Yao, M. C. and Gall, J. G.**, Alteration of the *Tetrahymena* genome during nuclear differentiation, *J. Protozool.*, 26, 10, 1979.
64. **Cummings, D. J.**, Studies on macronuclear DNA from *Paramecium aurelia, Chromosoma,* 53, 191, 1975.
65. **Colgin-Bukovsan, L. A.**, The genetics of mating types in the suctoria *Tokophrya lemnarum, Genet. Res.,* 27, 303, 1976.
66. **Colgin-Bukovsan, L. A.**, Life cycles and conditions for conjugation in the suctoria *Tokophyra lemnarum, Arch. Protistenk.*, 121, 223, 1979.
67. **Rudzinska, M. A.**, The influence of amount of food in the reproduction rate and longevity of suctorian *Tokophrya infusionum, Science,* 113, 10, 1951.
68. **Rudzinska, M. A.**, The differences between young and old organisms in *Tokophrya infusionum, J. Gerontol.*, 10, 469, 1955.
69. **Rudzinska, M. A.**, The use of a protozoan for studies on ageing. I. Differences between young and old organisms of *Tokophrya infusionum* as revealed by light and electron microscopy, *J. Gerontol.*, 16, 213, 1961a.
70. **Rudzinska, M. A.**, The use of a protozoan for studies on ageing. II. Macronucleus in young and old organisms of *Tokophrya infusionum:* light and electron microscopy, *J. Gerontol.*, 16, 326, 1961b.
71. **Rudzinska, M. A.**, The use of a protozoan for studies on ageing. III. Similarities between young overfed and old normally-fed *Tokophrya infusionum:* a light and electron microscope study, *Gerontologia,* 6, 206, 1962.
72. **Nanney, D. L.**, Inbreeding degeneration in *Tetrahymena, Genetics,* 42, 137, 1957.
73. **Allen, S. L.**, Cytogenetics of genomic exclusion in *Tetrahymena, Genetics,* 55, 797, 1967.
74. **Nanney, D. L.**, Aging and long-term temporal regulation in ciliated protozoa. A critical review, *Mech. Aging Dev.*, 3, 81, 1974.
75. **Nanney, D. L.**, Molecules and morphologies: the perpetuation of pattern in the ciliated protozoa, *J. Protozool.*, 24, 27, 1977.
76. **Simon, E. M. and Nanney, D. L.**, Germinal aging in *Tetrahymena thermophila, Mech. Ageing Dev.*, 11, 253, 1979.
77. **Bruns, P. J. and Brussard, T. E. B.**, Nullisomic *Tetrahymena:* elimination of germinal chromosomes, *Science,* 213, 549, 1981.
78. **Coleman, A.**, Sexuality in colonial green flagellates, in *Biochemistry and Physiology of Protozoa,* 2nd ed., Levandowsky, M. and Hutner, S. H., Eds., Academic Press, N.Y., 1979, 307.
79. **Weise, L.**, Genetic aspects of sexuality in *Volvocoles,* in *The Genetics of Algae,* Lewin, R. A., Ed., University of California Press, Berkeley, 1976, 174.
80. **Hagen, G. and Kochert, G.**, Protein synthesis in a new system for the study of senescence, *Exp. Cell Res.*, 127, 451, 1980.
81. **Kochert, G.**, Differentiation of reproductive cells in *Volvox carteri, J. Protozool.*, 15, 438, 1968.
82. **Starr, R. C.**, Structure, reproduction and differentiation in *Volvox carteri, f. nagariensis Iyengar,* strains *HK9 and 10, Arch. Protis.*, 111, 204, 1969.
83. **Pommerville, J. C. and Kochert, G. D.**, Changes in somatic cell structure during senescence of *Volvox carteri, Eur. J. Cell Biol.*, 24, 236, 1981.
84. **Sessoms, A. H. and Huskey, R. J.**, Genetic control of development in *Volvox:* isolation and characterization of morphogenetic mutants, *Proc. Natl. Acad. Sci. U.S.A.*, 70, 1335, 1973.
85. **Muggleton, A. and Danielli, J. F.**, Inheritance of the ''life-spanning'' phenomenon in *Amoeba proteus, Exp. Cell Res.*, 49, 116, 1968.
86. **Muggleton-Harris, A. L.**, Reassembly of cellular components for the study of aging and finite life span, *Int. Rev. Cytol.*, 279(Suppl.), 208, 1979.

87. **Widdus, R., Tayler, M., Powers, L., and Danielli, J. F.,** Characteristics of the "life spanning" phenomenon in *Amoeba proteus, Gerontologia,* 24, 208, 1978.
88. **Sousa e Silva, E.,** Ultrastructural variations of the nucleus in dinoflagellates throughout the life cycle, *Acta Protozool.,* 31, 227, 1977.
89. **Gomez, M. P., Harris, J. B., and Walne, P. L.,** Ultrastructural cytochemistry of *Euglena gracilia* Z: from aging cultures, *J. Protozool.,* 20, 515, 1973.
90. **Elliot, A. M. and Bak, I. J.,** The fate of mitochondria during aging in *Tetrahymena pyriformis, J. Cell Biol.,* 20, 113, 1964.
91. **Grusky, G. E. and Aaronson, S.,** Cytochemical changes in aging *Ochromonas:* evidence for an alkaline phosphatase, *J. Protozool.,* 16(4), 686, 1969.
92. **Fok, A. K. and Allen, R.D.,** Axenic *Paramecium caudatum.* II. Changes in fine structure with culture age, *Eur. J. Cell Biol.,* 25, 182, 1981.
93. **Fok, A. K., Allen, R. D., and Kaneshiro, E. S.,** Axenic *Paramecium caudatum.* III. Biochemical and physiological changes with culture age, *Eur. J. Cell Biol.,* 25, 193, 1981.
94. **Sonneborn, T. M. and Schneller, M. V.,** A genetic system for alternative stable characteristics in genomically identical homozygous clones, *Dev. Genetics,* 1, 21, 1979.
95. **Nanney, D. L.,** Calcium chloride effects on nuclear development in *Tetrahymena, Genet. Res.,* 27, 297, 1976.
96. **Nanney, D. L. and Meyer, E.B.,** Traumatic induction of early maturity in *Tetrahymena, Genetics,* 86, 103, 1977.
97. **Koizumi, S.,** The cytoplasmic factor that fixes macronuclear mating type determination in *Paramecium aurelia,* syngen 4, *Genetics,* 68(Suppl.), s34, 1971.
98. **Skaar, P. D.,** Past history and pattern of serotype transformation in *Paramecium aurelia, Exp. Cell Res.,* 10, 646, 1956.
99. **Sonneborn, T. M.,** The cytoplasm in heredity, *Heredity,* 4, 11, 1950.
100. **Sonneborn, T. M.,** The *Paramecium aurelia* complex of fourteen species, *Trans. Am. Microsc. Soc.,* 94, 155, 1975.
101. **Fukushima, S.,** Effect of X-irradiations on the clonal lifespan and fission rate in *Paramecium aurelia, Exp. Cell Res.,* 84, 267, 1974.
102. **Tixador, R., Richoilley, G., Monrozies, E., Planel, H., and Tap, G.,** Effects of very low doses of ionizing radiation on the clonal lifespan in *Paramecium tetraurelia, Int. J. Radiat. Biol.,* 39(1), 47, 1981.
103. **Collins, C. and Johnson, R.,** Novobiocin: an inhibitor of the repair of UV - induced but not X-ray induced damage in mammalian cells, *Nucleic Acids Res.,* 7(5), 1311, 1979.
104. **Sung, S. C.,** Effect of novobiocin on DNA-dependent DNA polymerases from developing rat brain, *Biochim. Biophys. Acta.,* 361, 115, 1974.
105. **Lipetz, P. D. and Smith-Sonneborn, J.,** Biology of Aging, Gordon Conference, Laconia, N.H. August, 1980.
106. **Smith-Sonneborn, J., Lipetz, P. D., and Stephens, R. E.,** *Paramecium* bioassay of longevity modulating agents, in *Intervention in the Aging Process; Basic Research, Clinical Screening, and Clinical Programs,* Vols. I, II, Regelson, W. and Sinex, F. M., Eds., Alan Liss, N.Y., 1982, 253.
107. **Smith-Sonneborn, J. Lipetz, P. D., and Stephens, R. E.,** Novobiocin inhibition of dark repair and longevity in *Paramecium,* in *DNA Repair and its Inhibition,* Collins, A., Downes, C. S., and Johnson, R. T., Eds., Nucleic Acids Res. Symp. Ser., IRL Press, Oxford, in press.
108. **Sonneborn, T. M.,** Breeding systems, reproductive methods, and species problems in protozoa, in *The Species Problem,* Mayer, E., Ed., American Association for Advancement of Science, Washington, D.C., 1957.
109. **Takagi, Y. and Yoshida, M.,** Clonal death coupled with the number of fissions in *Paramecium caudatum, J. Cell Sci.,* 41, 177, 1980.
110. **Jennings, H. S.,** *Paramecium bursaria:* life history. II. Age and death of clones in relation to the results of conjugation, *J. Exp. Zool.,* 96, 17, 1944a.
111. **Williams, D. B.,** Clonal aging in two species of *Spathidium* (Ciliophora: Gymnostomatids), *J. Protozool.,* 27(2), 212, 1980.
112. **Kosaka, T.,** Autogamy in fresh water *Euplotes woodruffi,* (Ciliata), *Zool., Mag.,* 79, 302, 1970.
113. **Kosaka, T.,** Effect of autogamy on clonal aging in *Euplotes woodruffi,* (Ciliata), *Zool., Mag.,* 81, 302, 1972.
114. **Heckmann, K.,** Age-dependent intraclonal conjugation in *Euplotes crassus, J. Exp. Zool.,* 165(2), 269, 1967.
115. **Katashima, R.,** Breeding systems in *Euplotes patella, Jpn. J. Zool.,* 13, 39, 1961.
116. **Siegel, R. W.,** Mating types in *Oxytricha* and the significance of mating type systems in ciliates, *Biol. Bull.,* 110, 352, 1956.
117. **Allen, S. L.,** Cytogenetics of genomic exclusion in *Tetrahymena, Genetics,* 55, 797, 1967.

118. **Pierson, B. F.,** The relation of mortality after endomixis to the prior endomatic interval in *Paramecium aurelia, Biol. Bull.,* 74, 235, 1938.

119. **Rodermel, S. R. and Smith-Sonneborn, J.,** Age correlated changes in expression of micronuclear damage and repair in *Paramecium tetraurelia, Genetics,* 87, 259, 1977.

120. **Fukushima, S.,** Clonal age and the proportion of defective progeny after autogamy in *Paramecium aurelia, Genetics,* 79, 377, 1975.

121. **Sonneborn, T. M. and Schneller, M. V.,** Measures of the rate and amount of aging on the cellular level, in *The Biology of Aging,* Strehler, B. L., Ed., Am. Inst. Biol. Sci. Symp. 6, Waverly Press, Baltimore, 290, 1960a.

122. **Siegel, R. W.,** Genetics of Ageing and the Life Cycle in Ciliates, *Symp. Soc. Exp. Biol.,* 21, 127, 1967.

123. **Heifetz, S. R.,** Ultrastructure and Dynamics of the Macronucleus of *Paramecium tetraurelia* During Aging, Ph.D. dissertation, University of Wyoming, Laramie, 1979.

124. **Sundararaman, V. and Cummings, D. J.,** Morphological changes in aging cell lines of *Paramecium aurelia.* III. The effects of emetine on polysome formation, *Mech. Ageing Dev.,* 6, 393, 1977.

125. **Tixador, R., Richoilley, G., and Planel, H.,** Radiosensitivity variation of *Paramecium aurelia* as a function of the clonal age, *J. Protozool.,* 19 (Suppl.), 73, 1972.

Chapter 2

# SENESCENCE IN *CAMPANULARIA FLEXUOSA* AND OTHER CNIDARIANS

## Mary Anne Brock

## TABLE OF CONTENTS

## I. INTRODUCTION

The Cnidaria (Coelenterata) are probably the most primitive Metazoa that exhibit senescence. Determinate lifespans have been reported for the individual hydranths or polyps of several colonial marine species such as *Campanularia flexuosa, Bougainvillia carolinensis* and *Clytia johnstoni* which are members of the class Hydrozoa.[1,2] By contrast, some representatives of the class Anthozoa have apparently indeterminate lifespans. Early and often-quoted evidence on anthozoans states that individuals of two species of sea anemones, *Actinia mesembryanthemum* and *Cereus pedunculatus,* lived in aquaria where they were continuously observed for 66 and for 80 to 90 years, respectively.[3,4] The original paper by Ashworth and Annandale[5] described the lack of obvious external changes during the time of observation but noted that "the old ones *(C. pedunculatus)* are much more strongly affected by unfavourable conditions than those which are more than thirty years younger, and also are longer in recovering when conditions become again favourable." The greatest change observed in the older anemones was in their fertility; the oldest individuals produced only a few young while younger anemones produced hundreds. Another specimen, *A. mesembryanthemum,* produced normal and vigorous young at the age of 50 years. It appeared to be in excellent health until it died, apparently of natural causes. Whether environmental conditions played a role in the natural deaths of the long-lived anemones is unknown. Recent data now have provided a clue to the anemones' indeterminate life. Autoradiography following perfusion of individual animals with tritiated thymidine showed proliferating cell populations in all tissues of *Haliplanella luciae.*[6] Such a wide distribution of cycling cells is unlike observations on other cnidarian polyps.

Analogies with the indeterminate life of the anemones exist in several species of *Hydra* which are noncolonial hydrozoans. There had been controversy for some time over whether *Hydra* exhibits senescence. In early experimental studies to affirm the "immortality" of *Hydra,* Brien[7] used the vital strain, Nile blue sulfate, to mark cells in a defined ring near the mouth or hypostome; these cells progressively moved to the base of the hydranth and were lost. Later studies using grafting techniques, radioactive labels, and marker dyes confirmed the continuous movement of ectodermal and gastrodermal tissue both distally and proximally from a region on the body just below the tentacles.[8-10] A broad distribution of mitotic cells over the body column also was observed,[11] and their progeny replaced tissues lost throughout the life of the individual. Comfort[3] called this "a case of indeterminate growth coexisting with a final specific size." It therefore seems clear that *Hydra* does not age in the manner of several colonial hydrozoans.

Two important criteria in the selection of invertebrate animals for aging studies are that individual lifespans are short, i.e., days or weeks rather than years, and that senescent changes are similar to those of mammals. In the most extensively studied species of colonial cnidarians, the 50% mortality point for their hydranths is reached in 1 to 2 weeks, depending on the ambient temperature and season of the year. Longevity can be modified by ambient temperature and intermittent feeding, as has been shown for other phyla of invertebrates. During this short lifespan, certain cell types such as neurosecretory and epitheliomuscular display senescent changes that are similar to those of long-lived muscle and nerve cells of mammals. The sections which follow will discuss the development, senescence, and regression or death of hydranths of the colonial species, with emphasis on the marine cnidarian, *Campanularia flexuosa,* and the endogenous annual and semilunar rhythmicity in hydranth development and longevity in this species.

## II. LIFE CYCLE AND MORPHOLOGY OF HYDRANTHS OF A COLONIAL CNIDARIAN

Colonial species of Cnidaria collected from the Atlantic Coast of the U.S. can be grown in the laboratory for several years using either natural or artificial sea water media. The methods for initiating the cultures and descriptions of growth patterns have been extensively described for *C. flexuosa* by Hammett[12] and then later for related cnidarian species by Berrill and Crowell.[13-16] The usual procedure for establishing a colony is to attach a small portion of a wild colony to a glass microscope slide with thread. In 1 to 2 days, the growing tip of the stolon attaches to the surface of the slide and new uprights arise from the stolon (Figure 1). Upward growth from the stolon produces the first hydranth of the upright, and successive apical growth zones produce internodes and then pedicels and hydranth buds which branch alternately. After a short embryonic development, the tentacles of the maturing hydranth extend around the hypostome and mouth, and the adult hydranths may eat (Figure 1). They are fed newly hatched nauplii of the brine shrimp, *Artemia*, daily. The colonies are grown in circulating, aerated sea water and in a short time reach the size shown in Figure 1D. Detailed procedures for maintaining the cultures and for charting them in order to remove hydranths of specific ages have been described previously.[17]

Survivorship curves for adult hydranths of three colonial species, *C. flexuosa, B. carolinensis*, and *C. johnstoni* were published some time ago and are shown in Figure 2.[2] Longevity data for hydranths of *C. johnstoni* were collected from colonies grown at 17°C at Woods Hole, Mass. during July; the other two species were grown in a constant 17°C and constant darkness (DD) environment in the laboratory in Baltimore, Md. during the months of March through May. The mean lifespan for *C. flexuosa* hydranths was 7.0 days, and this agrees remarkably well with that of 6.8 days reported earlier by Crowell.[1] It is evident from Figure 2B that there is a nearly constant probability of death for *C. johnstoni* hydranths, which suggests that mortality is not related to hydranth age. This difference in the survivorship curve has been discussed in relation to the growth habit of *C. johnstoni* by Strehler.[4] However, the *C. johnstoni* colonies were observed during the month of July, and as will be discussed in another section, marked seasonal decreases in longevity of *C. flexuosa* hydranths occurred during the summer months even though the colonies were grown at constant ambient temperature. Pertinent data may be seen in the hydranth survivorship curves during phases of minimum and maximum longevity of *C. flexuosa* cultured continuously at 24°C (Figure 16C). A similar endogenous rhythmicity in hydranth longevity may exist in other colonial cnidarians, and therefore it would be judicious to repeat the observations on *C. johnstoni* during other seasons of the year and possibly at other ambient temperatures before concluding that mortality in this species is not related to hydranth age.

The life of adult hydranths ends with regression. The process was carefully described for *Obelia geniculata* and *Campanularia sp.* by Huxley and de Beer[18] who observed that the hydranths gradually lost their form as cells dedifferentiated in the tentacles (see Figure 3 for illustration of stages in the regression of *C. flexuosa* hydranths). The tentacles shortened, and in later stages cells and debris derived from distal portions of the hydranth filled the digestive cavity and were propelled by pulsations from the regressing hydranth into the pedicel. The debris, which was called regression fragments by Lunger, is phagocytized by digestive cells of adult hydranths in the colony.[19] Regression is an autolytic process as will be discussed later, and the regression fragments resemble secondary lysosomes. The hydranth body gradually becomes smaller and an empty hydrotheca finally remains at the site of regression and is detached from the pedicel before the initiation of another hydranth bud at the same location. The cycle of regression and replacement of hydranths at successive locations on an upright has been described in detail by Crowell.[1]

The hydranths of *C. flexuosa* are radially symmetrical and have a single ring of tentacles that surround the hypostome, a distal projection with an opening to the coelenteron or

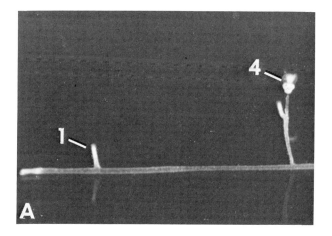

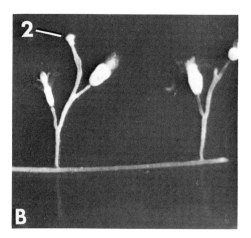

FIGURE 1.    Growth of a young colony of *C. flexuosa* from the initiation of new uprights near the growing tip of the stolon (A) to a later stage showing lateral stolon growth and the alternate branching pattern of older uprights (C, D). The differentiation of hydranths begins with growth of the pedicel (1) which enlarges to resemble a cone (2). Distal regions of the cone increase in diameter so it appears cylindrical in shape (3); initiation of a peripheral ring of tentacles on the cylinder and their differentiation results in the short-tentacle stage (4). The tentacles then extend to surround the central hypostome which has an opening to the coelenteron or gastrovascular cavity (see Figure 4). An *Artemia* being ingested is shown in (C) (arrow). The clear specks in the background in (D) are *Artemia* in the seawater medium; the distended appearance of the adult hydranths in (C) and (D) is due to the ingestion of several brine shrimp. An adult hydranth has regressed at the empty location in (C) where only the chitinous perisarc or hydrotheca remains. (4A—C, magnification × 20; 4D, magnification × 10.)

gastrovascular cavity (Figure 4). Two epithelia, the ectoderm and gastroderm, are separated from each other by an acellular mesoglea, which is a type of connective tissue. The gastroderm of the tentacles is a cylindrical column of highly vacuolated cells and is separated from the glandular gastrodermal cells of the body at the base of the tentacles by the mesoglea. The ultrastructure of the cell types listed in Figure 4 has been described for young hydranths and structural differences were noted in certain of the cell types of older hydranths.[20,21] A

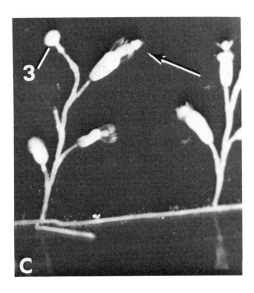

FIGURE 1

comprehensive review of cnidarian histology by Chapman includes electron micrographs of other representatives of the phylum.[22]

In older *C. flexuosa* hydranths, the structural changes displayed by cell types that are presumably long-lived include (1) conspicuous secondary lysosomes and residual bodies in the epitheliomuscular and neurosecretory cells of the body and tentacle ectoderm in the gastrodermal cells of the tentacles, and (2) large lipid droplets in the apical portions of digestive cells. Since these already have been described in detail,[21] representative cell types were chosen to illustrate the structural alterations in aging and regressing hydranths. Figures 5 to 8 show what may be interpreted as progressive alterations in the hydranth tentacles. The typical structure which accounts for the ladder-like appearance of the tentacles is shown in a section from a 3-day-old hydranth (Figure 5). As the hydranths aged, secondary ly-

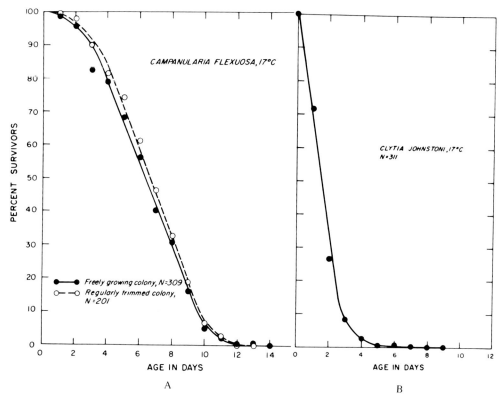

FIGURE 2.    Survivorship of the hydranths of *Campanularia flexuosa, Bougainvillia carolinensis,* and *Clytia johnstoni. C. flexuosa* and *B. carolinensis* colonies were grown in aerated, circulating, artificial seawater in Baltimore, Md. *C. johnstoni* colonies were grown at Woods Hole, Mass. in circulating seawater. All were kept at an ambient temperature of 17°C. (From Brock, M. A. and Strehler, B. L., *J. Gerontol.,* 18, 23, 1963. With permission.)

sosomes and residual bodies were commonly observed in both the epitheliomuscular and gastrodermal cells of the tentacles (Figures 6 to 8). The presence of these is considered evidence for the autolysis of cellular components, and cytochemical tests indicated that many displayed positive reactions for acid phosphatase activity.[26] Abundant autophagic vacuoles filled the vacuolar spaces in the epitheliomuscular cells, and small cytoplasmic bodies occupied intercellular spaces as is apparent in Figure 7. When regression begins, changes in hydranth form first appear in the tentacles as they become shorter, a possible result of collapse of gastrodermal cell vacuoles. The compact appearance of these cells is shown in Figure 8.

The basal portions of digestive cells in the gastroderm of the young hydranth body (Figure 9) also undergo marked alterations with age as exemplified by the appearance of prominent residual bodies and lipid droplets in a 9-day-old hydranth (Figure 10). The autolytic process eventually involves most of the recognizable body cell types (Figures 11 and 12), and the resulting membrane-bounded structures resemble the regression fragments which are phagocytized by digestive cells of adult hydranths as described by Lunger.[19] In early light microscopic studies, Huxley and de Beer noted that "small masses of dense pigment are often found in the partly resorbed zooid, representing products of degeneration."[18]

Although postmitotic cells such as cardiac, muscle, and neurones of older mammals also display lysosomal structures similar to those of older *C. flexuosa* hydranths, activation of the hydrolytic enzymes of lysosomes is not confined to postmitotic cell types but is a prominent feature in the development and remodeling of a variety of tissues.[27,28] The stimuli which initiate autolytic activity in older cells are not well understood. For this reason there

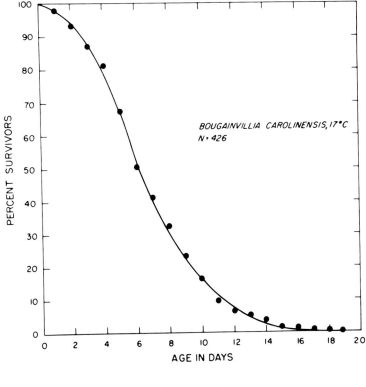

FIGURE 2C.

are advantages in utilizing short-lived organisms such as *C. flexuosa* to investigate whether injury to cell organelles or other intracellular events precipitate the activation of lysosomal enzymes.

As adjuncts to transmission electron microscopy, other ultrastructural techniques such as scanning electron microscopy, autoradiography, freeze-fracturing, cytochemistry, and electron probe microanalysis have recently been used in analyzing, for example,

1. The three-dimensional structure of *Hydra attenuata* epitheliomuscular cells and their relation to morphogenesis[29]
2. The fine structure of septate and gap junctions of *Chlorohydra viridissima*[30]
3. The secretory processes involved in crosslinking skeletal collagen of *Leptogorgia virgulata*[31]
4. The chemical composition of the components of *Pennaria tiarella* nematocysts[32]

Such techniques may be of great value for studies of cellular senescence in the colonial cnidarian species which have finite lifespans.

## III. HYDROLYTIC ENZYMATIC ACTIVITY IN AGING HYDRANTHS

Changes in the activity of lysosomal enzymes can be documented during the life span of adult *C. flexuosa* hydranths, and these are associated with the cytological observations of increased autolysis. Histochemical methods for the demonstration of nonspecific acid phosphatase activity have provided semiquantitative data for its enhancement with age in ectodermal and gastrodermal cells of the tentacles and hydranth body (Figure 13). Hydranths of specific ages were fixed in formol-calcium and then incubated in a medium containing the

FIGURE 3.    Regression or death of adult hydranths of *C. flexuosa* is signaled by gradual shortening of the tentacles (1 and 2), so that eventually the hypostome projects distally (3). Further autolysis of the hypostome (4) and hydranth body obliterates the hydranth form (5), and an empty hydrotheca is left at the site of regression. (Magnification × 20.) (From Brock, M. A., *Comp. Biochem. Physiol.*, 51A, 377, 1975. With permission.)

substrate sodium β-glycerophosphate, following the modified Gomori method of Pearse.[33] Phosphate ions released by enzymatic hydrolysis from the substrate are precipitated by lead in the incubation medium, and ammonium sulfide is used to convert the lead phosphate to lead sulfide. The greatest mean increases in black, precipitated grains counted in paraffin sections were from 9.8 to 20.7 and from 28.7 to 72.4 gr/1000μ² in the body ectoderm and gastroderm, respectively. Some enzymatic activity is lost during fixation of the tissue; for example, in later observations there was a 50% reduction after fixation of *C. flexuosa* hydranths for $1^{1}/_{2}$ hr with 3% glutaraldehyde in 0.1 $M$ cacodylate buffer, containing 0.33 $M$ sucrose at pH 7.4 and 0°C. The histochemical methods, therefore, will provide data on relative enzymatic activity and its localization to specific epithelia or cell types.

Quantitation of nonspecific acid phosphomonoesterase in homogenates of *C. flexuosa* hydranths provided evidence for increased activity in 10- to 11-day-old hydranths compared to 2-day-old hydranths, determined 1 day after feeding the colonies. However, if the enzymatic activity was related to the DNA content per hydranth, the age-related increase was obscured (Table 1). For these assays, hydranths were homogenized gently in 0.33 $M$ sucrose at 4°C to prevent the rupture of primary lysosomal membranes and the release of latent enzymatic activity. The homogenates were incubated with sodium-β-glycerophosphate, with and without the addition of Triton X-100 which releases the bound lysosomal enzymes. Trichloroacetic acid was added at the end of the incubation period and, after centrifugation, inorganic phosphorus and DNA were determined in the supernatant and precipitate, respectively. The detailed procedures are given in the legend for Table 1. Since it was clear from the ultrastructural observations that increased activation of the lysosomal system was as-

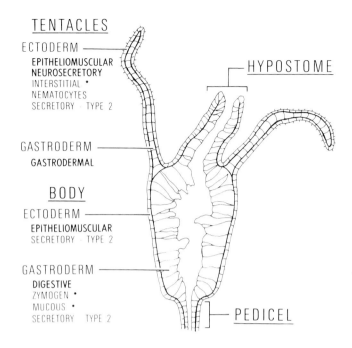

FIGURE 4. The cell types identified in the ectoderm and gastroderm of *C. flexuosa* hydranths are listed in this drawing made from a light micrograph. The bolder type identifies the cell types in which prominent age-associated ultrastructural changes were displayed. Less evident alterations in older hydranths were seen in the cell types marked with asterisks. The heavy line between the two epithelial layers represents the acellular mesoglea. (From Brock, M. A., *J. Ultrastr. Res.*, 32, 118, 1970. With permission.)

sociated with hydranth aging, the lack of change with age when enzymatic activity was based on DNA content per hydranth was puzzling. However, several cell types may continue to proliferate throughout the life of the hydranths to replace short-lived cells[36] and especially to enlarge the population of nematocytes which fire and lose their encapsulated nematocysts each time the colonies are fed. This could account for the rise in DNA per hydranth with age. The methods described here and also the recent polyacrylamide gel electrophoresis assays of acid and alkaline phosphatase and other enzymes to clarify taxonomic relationships of several cnidarian genera[37] show the applicability of microdeterminations for study of the aging process in small invertebrates.

## IV. METABOLIC AND FUNCTIONAL MEASUREMENTS

### A. Metabolism

The accumulation of large lipid droplets in the apices of older digestive cells is one of the few indications that changes in intermediary metabolism may be associated with senescence of *C. flexuosa* hydranths. In addition, 300 to 400 Å particles thought to represent glycogen were commonly seen in the mesoglea of young but not old hydranths.[21] Since the mesoglea serves in part to transport metabolites from gastrodermal digestive cells to the ectoderm, the near absence of glycogen suggested a diversion in the metabolism of ingested food that resulted in storage of lipid in droplets. Some time ago, Strehler and Crowell[38] reported that the ATP content per hydranth fell in the 4th and 5th days of life to $1/4$ of the levels observed in 1-day-old individuals.[38] Since their calculations used the hydranth as a reference either changes in intracellular ATP concentration or cell number could account

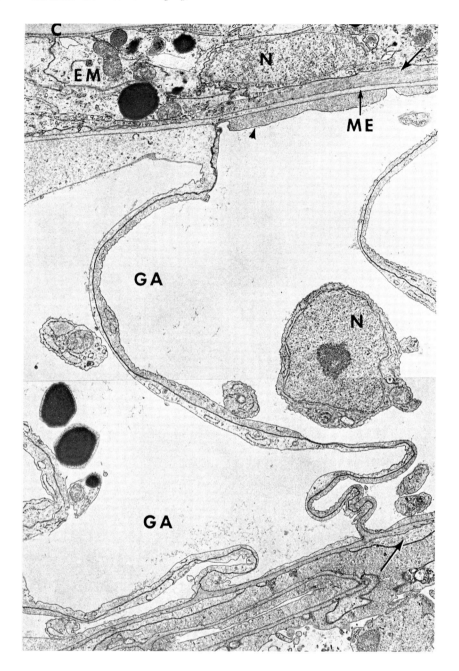

FIGURE 5. Epitheliomuscular cells of the tentacle ectoderm (EM) in longitudinal section of a 3-day-old hydranth have an eccentric nucleus (N), and their interdigitating basal processes containing myofibrils (arrows) rest on the acellular mesoglea (ME). Cytoplasmic inclusions are mitochondria, lipid droplets, vacuoles of varying size, and a few structures which resemble small secondary lysosomes. The cellular surface is covered by the cuticle (C). The highly vacuolated gastrodermal cells (GA) have a narrow band of cytoplasm at the periphery of the cells which forms the septa between cells and is adjacent to the mesoglea. Circularly oriented myofibrils are often seen in cross-section in the cytoplasm near the mesoglea (arrow head). The nucleus and other cellular components may appear to be suspended in the central vacuole if cytoplasmic projections to the periphery of the cell are in another plane of section. Dilated cisternae of the rough endoplasmic reticulum (RER) are seen in the septa. (magnification × 8700.) Hydranths used for Figures 5, 6, and 7 were fixed for 2 hr at 0 to 4°C in 3% glutaraldehyde in 0.075 *M* cacodylate buffer, pH 7.4[23] which contained 0.33 *M* sucrose. After thorough washing, they were post-fixed for 1 hr at 0 to 4°C in 1% $OsO_4$ in 0.06 *M* phosphate buffer, pH 7.3, containing 0.016 *M* sucrose.[24] Thin sections of material embedded in Epon were stained with either lead citrate or uranyl acetate.

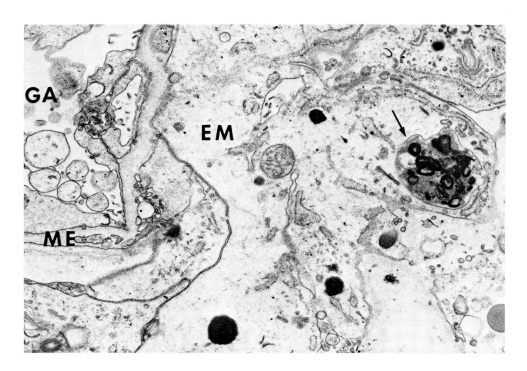

FIGURE 6.   An epitheliomuscular cell (EM) in the tentacle ectoderm of an 8-day-old hydranth. The residual body (arrow) contains electron-dense myelin figures. Otherwise the cell type appears similar to younger hydranths. Isolated cytoplasmic fragments that resemble autophagic vacuoles appear in the vacuole of a gastrodermal cell (GA) near the mesoglea (ME). (Magnification × 14,000.)

for the decline in ATP. Although DNA content per hydranth rose (Table 1) and total protein per hydranth remained relatively constant (Table 2) with age, neither observation allows the conclusion that any specific cell types exhibit a marked decline in ATP content.

Since cnidarians are animals at the tissue level of organization, the biochemical determination are comparable to those made on homogenized tissues with several cell types in more complex animals. Therefore, they lack the specificity desired for assessing possible senescent changes in single cell types. There are possibilities however, for more discriminating measurements with the use of techniques for the quantitative analysis of cell types described for *Hydra attenuata* by David[40] and then applied to the analysis of changes in cell distributions during *Hydra attenuata* morphogenesis.[41] Cnidarians offer the advantage of having only a few cell types which are all recognizable in the single cell suspensions resulting from the use of these techniques. The method utilizes a glycerin-glacial acetic acid-water mixture (1:1:13) to dissociate the cells which are then fixed. David's paper[40] discussed advantages in application of the methods for the analysis of cellular components of tissues, kinetic experiments on cellular turnover, autoradiography, and quantitative microspectrophotometry. Later, the maceration technique was used to provide single cell types from *Hydra attenuata* or *Chlorohydra viridissima* in studies as diverse as (1) epithelial and interstitial cell cycle kinetics and nerve and nematocyte differentiation,[42-44] (2) the influence of cell density on stem cell differentiation to nerve cells and nematocytes,[45] (3) quantitation of phagocytic activity of digestive cells,[46] and (4) the effects of different feeding regimens on relative sizes of cell populations.[47] It seems feasible that such quantitative analyses also could be applied to the study of specific cell types during the aging of hydranths in colonial species.

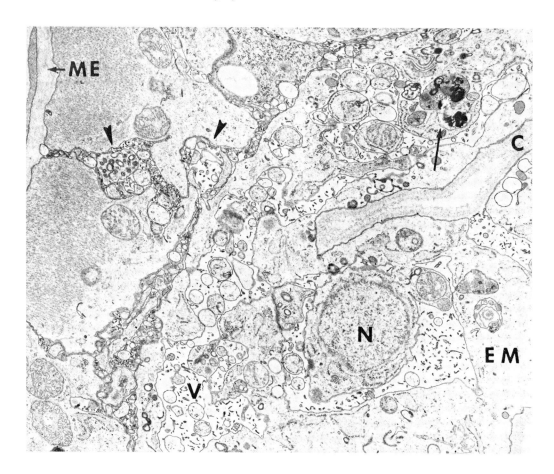

FIGURE 7.   A residual body containing electron-dense myelin figures in an amorphous matrix (arrow) is in a cytoplasmic projection into the vacuole of this epitheliomuscular cell (EM) from the tentacle of an 8-day-old hydranth. Other projections and isolated cytoplasmic fragments fill the vacuole (V). These resemble autophagic vacuoles since they contain unaltered cytoplasmic organelles and are single-membrane bounded. Conspicuous myofibrils lie in the basal processes adjacent to the mesoglea (ME), and numerous small cytoplasmic bodies fill intercellular spaces between the basal processes (arrow heads). (N, nucleus; C, cuticle.) (Magnification × 8700.)

## B. Function

A serious inquiry about senescent changes in the functional capacity of cnidarians or their cell types has yet to be made. The feeding ability and rates for ingestion and digestion of brine shrimp did not differ in hydranths from 1 to 7 or 8 days of age.[38] Ths was taken as evidence that regression cycles had not evolved as a mechanism for replacing inefficient parts. As the ultrastructural picture shows, brine shrimp are phagocytized and digested in apical areas of digestive cells, so the observations on food digestion[38] suggest a lack of change with age in the function of plasma membranes and phagocytic vacuoles of the digestive cells which line the coelenteron. The fate of the metabolites and functional capacity of other cell types are unknown except for the inference that nematocyte function may not be impaired with age. Nematocytes contain the nematocysts, projectiles which are stimulated by the brine shrimp to fire and immobilize them.[48] Whether the number of nematocytes stimulated per brine shrimp changes with age remains to be explored.

The studies of Morin and Cooke[49-51] on the physiology of the colonial cnidarian, *Obelia geniculata*, which is in the same family as *C. flexuosa,* revealed that spontaneous electrical potentials were generated in unstimulated hydranths. Their meticulous analysis showed that

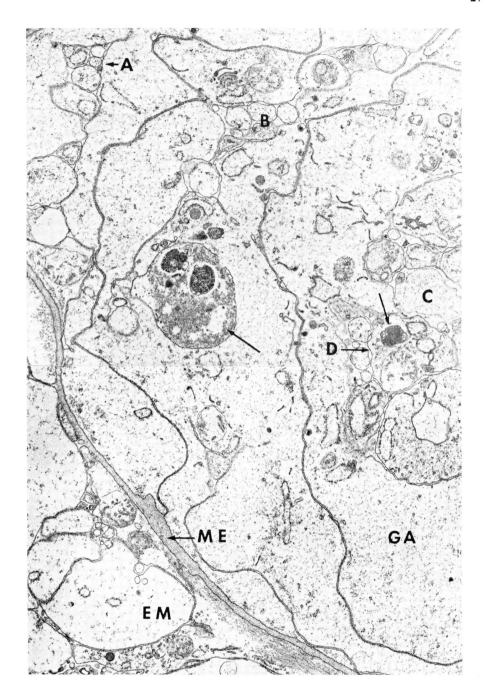

FIGURE 8. Gastrodermal cells (GA) in distal portions of the tentacles of regressing hydranths no longer display the large central vacuoles seen in Figure 5. The apparently collapsed cells contain residual bodies (arrows) and what may be interpreted as cytoplasmic bodies or tubular extensions from adjacent cells filling intercellular spaces (A). Autophagic vacuoles fill the collapsed central vacuoles (B and C); one nearly segregated cellular projection (D) contains degraded material as well as unaltered cellular organelles. The myofibrils in the basal processes of the epitheliomuscular cells (EM) adjacent to the mesoglea (ME) appear structurally unaltered. (Magnification × 15,000.) Hydranths used for Figures 8 to 12 were fixed for 1 1/2 to 2 hr at 0 to 4°C in 6% glutaraldehyde in 0.075 $M$ cacodylate buffer, pH 7.6.[23] After thorough washing, they were post-fixed for 1 1/2 hr at 0 to 4°C in 1% $OsO_4$ in Veronal acetate buffer, pH 7.4, containing 0.33 $M$ sucrose.[25] Epon-embedded thin sections were stained with either lead citrate or uranyl acetate.

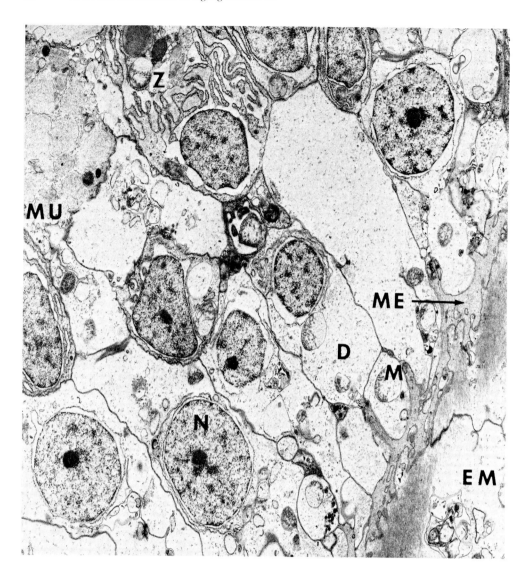

FIGURE 9.    The basal portions of several digestive cells (D) in the body gastroderm of a 2-day-old hydranth have scattered mitochondria (M) and cisternae of the RER in their homogeneous-appearing cytoplasm. Large membrane-bounded mucous droplets pack the mucous cell (MU) in the upper left. The RER of the zymogen cell (Z) is highly dilated and two zymogen droplets appear in this section. Longitudinally oriented myofibrils in the interdigitating bases of the ectodermal epitheliomuscular cells (EM) lie adjacent to the mesoglea (ME). (N, nucleus.) (Magnification × 8700.)

three behavioral patterns, hydranth contraction, mouth opening, and individual tentacle flexion, were associated with their respective contraction, mouth opening, and tentacle contraction potentials. Electrical activity and bioluminescence in adjacent hydranths following mechanical, electrical, and chemical stimulation were also documented in *O. genicu-lata.*[50,51] Because of the amplitude and duration of the potentials, they were thought to be epithelial in nature and not to originate from neurons. Even though these measurements on individuals as small as the adult hydranths of colonial cnidarians require extreme care, their application to the study of the behavioral physiology of aging hydranths could prove fruitful.

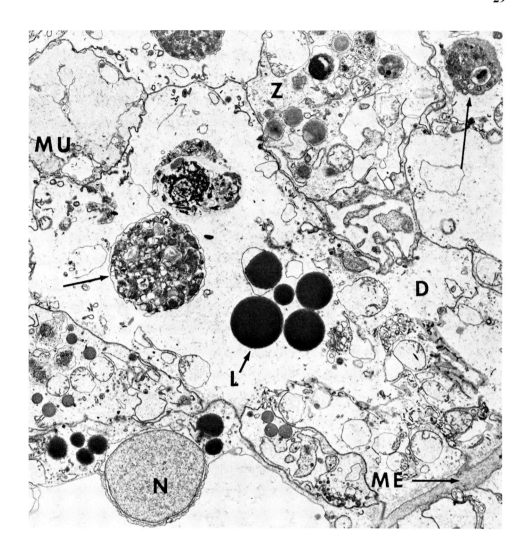

FIGURE 10. Conspicuous residual bodies (arrows) were common in basal portions of gastrodermal digestive cells (D) in older animals, such as these from a 9-day-old hydranth. The cytoplasm included large lipid droplets (L), mitochondria, and scattered elements of the RER. Several mucous droplets (MU) have apparently coalesced in the mucous cell. The highly dilated RER and zymogen droplets resemble those of zymogen cells (Z) from young hydranths. (N, nucleus; ME, mesoglea.) (Magnification × 7800.)

## V. ENDOGENOUS RHYTHMICITY AND ENVIRONMENTAL MODIFICATION OF THE LONGEVITY OF *C. FLEXUOSA* HYDRANTHS

Circannual and circasemilunar rhythmicity in the development, growth, and longevity of *C. flexuosa* hydranths were observed in colonies of a single clone grown under constant conditions at 10, 17, and 24°C.[17,52-56] These endogenous rhythms satisfied the requirements given by Aschoff et al.[57] that the "system is capable of self-sustained oscillations under constant conditions." Such free-running rhythms persist in the absence of known exogenous signals or Zeitgebers from the environment and are temperature-compensated,[58,59] (i.e., the periodicity of the oscillations remains relatively uniform at different ambient or body temperatures) but with time the oscillations gradually lose synchrony with environmental cycles.

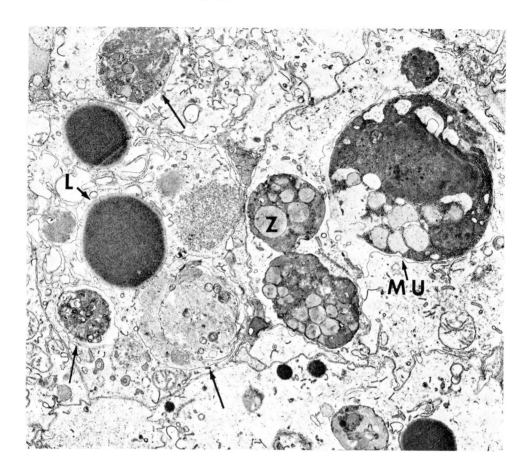

FIGURE 11.   The basal portions of gastrodermal cells of regressing hydranths contain conspicuous secondary lysosomes and residual bodies (arrows). Recognizable mucous (MU) and zymogen droplets (Z) appear in the homogeneous, dense, apparently degraded matrix of the secondary lysosomes. Myelin whorls, membranous residues, and electron-dense amorphous material fill the residual bodies. (L, lipid droplet.) (Magnification × 15,000.)

## A. Circannual Rhythmicity

The marked seasonal differences in the growth of *C. flexuosa* colonies are readily apparent in Figure 14. Changes in developmental patterns and rates of hydranth budding contributed to the interruptions of the luxuriant growth habit and the reversals from sparse to luxuriant growth during two phases of the circannual cycle. Figure 15 shows the periodicity of the reversals in growth habit of colonies grown at 10, 17, and 24°C. The detailed culture methods have been described,[17] and it was assumed that Zeitgebers were absent since (1) the free-running periods were longer than 365 days, and (2) colonies obtained from their natural habitat at Woods Hole, Mass. in November 1971 were out of phase with those which had been cultured continuously in the laboratory since May 1968 (Figure 15).

The clear alterations in growth habit during the phases of sparse growth were associated with at least three developmental changes: (1) an increase in the percentage of hydranth buds that abort rather than undergo development to normal adulthood, (2) a decline in the rate of initiation of hydranth buds in apical growth zones and the virtual absence of hydranth replacement at sites of regression, and (3) the death of apical portions of uprights where replacement of regressed hydranths had not taken place.[17,52] Most surprisingly, a difference in the survivorship of adult hydranths in the colonies occurred in concert with the developmental cycles despite their culture at constant ambient temperatures.[53] Figure 16 shows

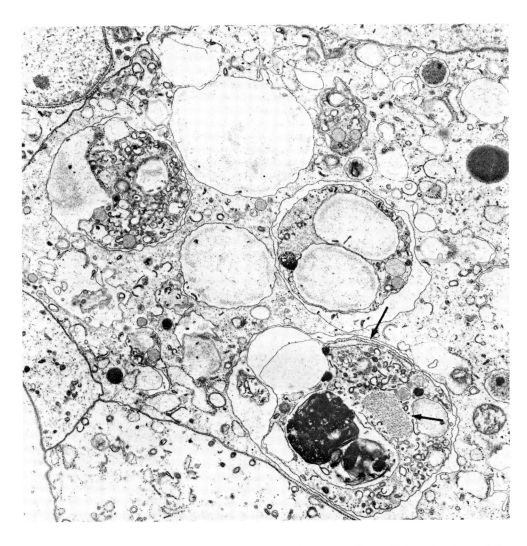

FIGURE 12.  Partially degraded cellular components, unrecognizable except for myofibrils sectioned tangentially, and elements of the RER (arrows), fill the secondary lysosomes and residual bodies in this regressing hydranth. (Magnification × 14,000.)

that longevity of the hydranths doubled during the phases of luxuriant growth. The rhythmicity in longevity which persisted for over the 2 years that daily observations were made and the yearly changes in coastal water temperature at the native environment of *C. flexuosa* (Woods Hole, Mass.) are illustrated in Figure 17. The endogenous rhythms of the colonies grown at three different ambient temperatures were well synchronized with each other and were inversely related to the water temperature at Woods Hole, except in January and February of each year when the most curtailed growth was observed. This as yet unexplained decline has analogies with the winter depression in physiological functions of mammals.[17] It is interesting that the endogenous changes in the rate of aging preceded the related cycles in the water temperature, that is, the increasingly longer lifespans in July 1969 occurred before the decline in water temperature. This anticipation of daily or seasonal cycles has selective advantage in preparing organisms for ensuing environmental conditions.[60,61]

The expected inverse effect of environmental temperature on the longevity of poikilothermic animals[3] was not clear during all seasons in the colonies grown at 17 and 24°C until their 2nd year in constant environments. This is shown in the near coincidence of the peaks

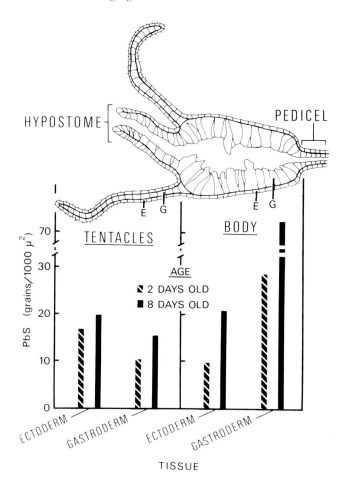

FIGURE 13.    The modified Gomori histochemical method to demonstrate acid phosphatase activity shows greater numbers of precipitated lead sulfide grains in the ectoderm (E) and gastroderm (G) of both the tentacles and body of 8-day-old *C. flexuosa* hydranths as compared with 2-day-old hydranths. They were fixed in formol-calcium for 1 1/2 hr, washed, and then incubated at 17°C for 5 hr in 0.01 $M$ sodium β-glycerophosphate in 0.05 $M$ acetate buffer (pH 5.0) containing 0.004 $M$ lead nitrate. The material was then immersed briefly in ammonium sulfide, dehydrated, paraffin embedded, and sectioned. The increase in the number of lead sulfide grains per 1000 $\mu^2$ was highly significant in the body epithelia ($p < 0.01$ using Fisher's test of significance for small samples).

in longevity in December 1968 and November 1969 (Figure 17). In contrast, the colonies grown at 10°C exhibited rhythmic but distinctly longer hydranth lifespans throughout the observations. The effect of higher ambient temperatures on hydranth survivorship in November 1970 is shown in Figure 18. The differences in the initial effectiveness of ambient temperature on longevity surely indicate that there are complex and poorly understood interactions between temperature and endogenous rhythmicity in hydranth longevity. Why there is not a constant, immediate effect on the colonies grown at 24°C is a mystery.

## B. Circasemilunar Rhythmicity

The circannual rhythms of *C. flexuosa* were comprised of continuously repeating rhythms with a periodicity close to lunar or semilunar. To illustrate these, records of hydranth longevity on each successive day in colonies freerunning at 10°C since May 1968 are shown

## Table 1
## ACID PHOSPHOMONOESTERASE RELEASE OF INORGANIC PHOSPHORUS AND DNA CONTENT OF HYDRANTHS IN *CAMPANULARIA FLEXUOSA* COLONIES CULTURED AT 10°C

### Inorganic Phosphorus Released in 60 min at pH 4.6

| Hydranth age (days) | N | $\mu M \times 10^{-3}$/hydranth (Mean ± S.E.) | |
|---|---|---|---|
| | | Free | Total |
| 2 | 6 | 0.93 ± 0.13 | 1.56 ± 0.13 |
| 10—11 | 6 | 1.83 ± 0.20[a] | 2.71 ± 0.33[a] |

### DNA Content

| Hydranth age (days) | N | $\mu g$ DNA $\times 10^{-3}$/ hydranth (Mean ± S.E.) |
|---|---|---|
| 2 | 6 | 5.25 ± 0.55 |
| 10—11 | 6 | 8.12 ± 0.87[b] |

### Inorganic Phosphorus Released ($\mu M \times 10^{-3}$)/DNA ($\mu g \times 10^{-3}$)/Hydranth (Mean ± S.E.)

| Hydranth age (days) | N | Free | Total |
|---|---|---|---|
| 2 | 6 | 0.175 ± 0.01 | 0.304 ± 0.02 |
| 10—11 | 6 | 0.237 ± 0.04 | 0.350 ± 0.06 |

*Note:* Acid β-glycerophosphatase activity was determined in *C. flexuosa* hydranths of specific ages by incubating homogenates with 0.01 *M* sodium β-glycerophosphate in 0.05 *M* acetate buffer, pH 4.6, containing 0.33 *M* sucrose for 60 min. at 10°C. Triton® X-100 at a final concentration of 0.143% was added to appropriate homogenates for the determination of total enzymatic activity. The optimal pH for the incubation of hydranths grown at 10°C was determined by using a range of acetic acid concentrations in the buffer for a series of homogenates incubated at 10°C. The incubation was terminated by adding an equal volume of 10% TCA. After centrifugation, inorganic phosphorus was assayed on 0.5 mℓ of the supernatants with the methods of Chen et al.[34] and DNA determinations were made on the precipitate after lipid extraction using the fluorometric procedures of Kissane and Robins.[35]

The increases in the free and total inorganic phosphorus and DNA of the older hydranths were highly significant using Fisher's test of significance for small samples.

The assays were done during the months of July and August 1970. In this phase of the circannual rhythm of hydranth longevity, 50% mortality was reached in 9 to 10 days (see Figure 16).

[a] $p < 0.01$
[b] $p < 0.02$

**Table 2**
**TOTAL PROTEIN OF**
**HYDRANTHS IN** *CAMPANULARIA*
*FLEXUOSA* **COLONIES**
**CULTURED AT 17°C**

| Hydranth age (days) | N | μg protein/hydranth (mean ± S. E.) |
|---|---|---|
| 1 | 6 | 2.02 ± 0.13 |
| 2 | 3 | 1.68 ± 0.07 |
| 3 | 4 | 2.18 ± 0.14 |
| 4 | 5 | 1.87 ± 0.05 |
| 5 | 5 | 2.24 ± 0.23 |
| 6 | 4 | 2.05 ± 0.09 |
| 7 | 6 | 2.32 ± 0.19 |
| 8 | 5 | 2.52 ± 0.32 |

*Note:* Protein content of specific aged hydranths was determined using the procedures of Lowry et al.[39] The assays were done during the month of August 1968. In this phase of the circannual rhythm of hydranth longevity, 50% mortality is reached in 4 to 5 days, and less than 5% of the adult hydranths live for 8 days (see Figure 16).

for February and March 1970 (Figure 19). During these months, the reversal from very sparse growth to the luxuriant growth phase occurred and corresponded in timing to the sharp increase in hydranth mean lifespans on February 22. The first peak in hydranth longevity shown in Figure 19 was just prior to new moon in February and hydranth lifespans were as long as 17 days then. They became progressively shorter on each successive day until the marked increase occurred on February 22, 1 day after full moon. For comparison, circasemilunar rhythmicity in hydranth regression, development, and longevity and their relationships to each other are illustrated for colonies 1 month after they were obtained from their native environment at Woods Hole (Figure 20). Successive peaks in hydranth regression, development, and longevity were observed in these colonies free-running at 17°C. The gradual change in the phase relationships of the peaks with the natural lunar cycle suggests that the rhythmicity is endogenous.

In other studies, colonies were transferred to higher and lower constant ambient temperatures during different phases of the circannual rhythm.[55] The effects of these temperature steps on longevity provided evidence for not only differential sensitivity to ambient temperatures during different phases of the circannual rhythm, but also phase-shifting of the rhythms of shorter periodicity. That is, peaks in hydranth longevity had a greater coincidence with full and new moon phases of the lunar cycle in colonies subjected to temperature steps than in control colonies free-running at one constant ambient temperature.

## C. Intermittent Feeding

The specific age of many poikilothermic invertebrates can be altered by restriction of food as well as changes in ambient temperature.[3] Intermittent feeding of *C. flexuosa* colonies at 10°C for 2 months increased the mean lifespan of the hydranths from 16.4 days for the controls which were fed daily to 21.0 days for those fed every 3rd day. Figure 21 shows survivorship curves for the two groups. The average number of brine shrimp ingested per hydranth was similar in both groups (see Figure 21 caption), and therefore it may be assumed

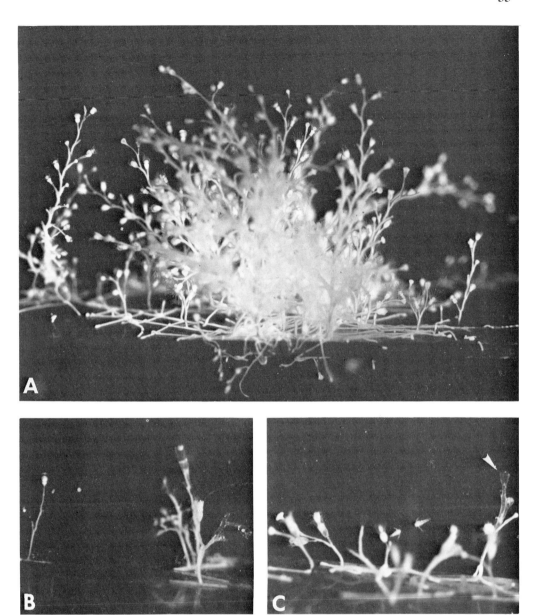

FIGURE 14. Luxuriant growth of the *C. flexuosa* colonies persisted during most of the circannual cycle (A). These colonies were cultured at 10°C, DD, and photographed in September 1969. In phases of sparse growth, changes in patterns and rates of development reduced the colonies to minimal size. These colonies were grown at 24°C (B) and 10°C (C), DD, and photographed during midwinter and summer phases, respectively. The spontaneous reversal to a phase of luxuriant growth is evident in (C) where empty uprights (arrow) remain along with young uprights similar to those in Figure 1. The periodicity of the reversals is shown in Figure 15. (14a, magnification × 5; 14b-c, magnification × 10.) (Figure 14A is from Brock, M. A., *Comp. Biochem. Physiol.*, 51A, 377, 1975. With permission.)

that the food consumption of intermittently fed hydranths was about one third that of the controls. The specific stage in the life cycle of the hydranths which is affected by the change in diet cannot be precisely identified since nutrients circulate throughout the colonies from the feeding adults to developing hydranth buds, uprights, and stolons. Nevertheless, the increased longevity associated with intermittent feeding parallels observations on other poikilothermic species and mammals.

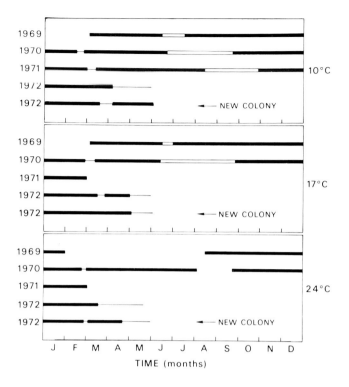

FIGURE 15.   *C. flexuosa* colonies cultured in constant environments at 10, 17, and 24°C grew luxuriantly during most of the circannual cycle (solid bars). Reversals to summer phases of sparse growth are indicated by the open bars, and the very sharply curtailed growth during midwinter phases is shown as narrow lines. The reversals in growth habit exhibited periodicities of greater than 365 days. Breaks in the records of colonies cultured at 17 and 24°C signify colony death, and other colonies were transferred from 10 to 17°C and then 24°C in order to complete the records through May 1972. The growth of colonies obtained from Woods Hole, Mass. in November 1971 is shown at the bottom of each panel. They were out of phase with the colonies cultured in the laboratory for almost 4 years. The inverse effect of environmental temperature on colonial longevity is illustrated by their increasing life spans, from about 12 months at 24°C to 24 months at 17°C and over 39 months at 10°C. The colonies cultured in the constant 10°C environment were very healthy when all observations were terminated in July 1972. (From Brock, M. A., *Comp. Biochem. Physiol.*, 51A, 385, 1975. With permission.)

Records of hydranth longevity for each successive day, similar to those illustrated in Figure 19, pinpointed the date that dietary manipulation effected the hydranth lifespans. Although intermittent feeding began on September 29, 1969, lifespans considerably longer than those of control hydranths were not observed until October 19. The peak in longevity was on October 24, 1 day before full moon; the mean lifespan was 34.5 days as compared to 19.5 days for the control hydranths on that date.[26] This delayed effect of intermittent feeding could mean that nutrient levels in the entire colony must be greatly reduced, and at least 20 days are required before any effect on longevity is seen. Alternatively, hydranths may be more sensitive to dietary manipulation during specific phases of the circasemilunar rhythm. A differential sensitivity seems plausible since the effects on longevity after perturbation by temperature steps were nearly coincidental with the new and full moon phases of the lunar cycle.[55]

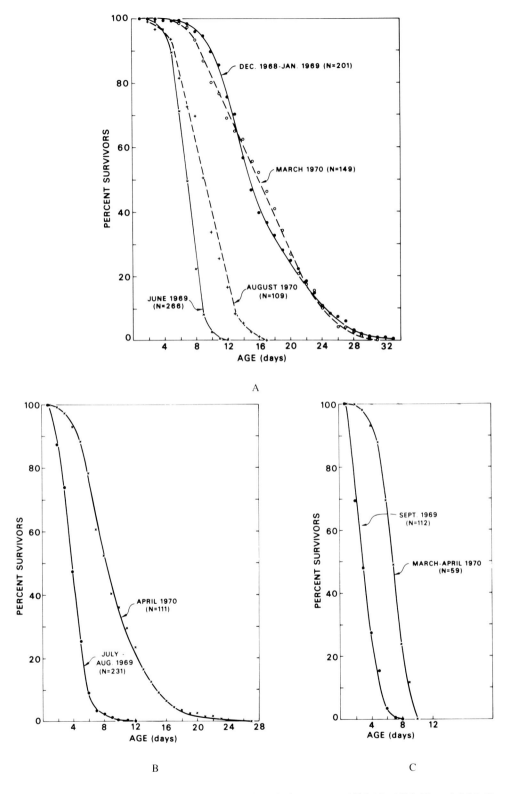

FIGURE 16. The lifespans of *C. flexuosa* hydranths in colonies grown at 10°C (A), 17°C (B), and 24°C (C), DD, were doubled in phases of luxuriant growth as compared to the phases of curtailed growth. The months in which observations were made are shown for the animals cultured continuously in each of the constant environments. (From Brock, M. A., *Comp. Biochem. Physiol.,* 51A, 391, 1975. With permission.)

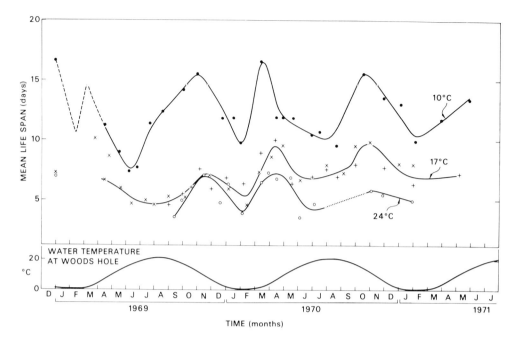

FIGURE 17.   Circannual rhythmicity in the longevity of hydranths in *C. flexuosa* colonies cultured in constant 10, 17, and 24°C environments persisted for over 2 years. The dashed line at the beginning of the record for colonies at 10°C indicates that there was a break in the observations made in the initial studies. Longevity data from the new colonies obtained from Woods Hole, Mass. in 1971 were used for January through March 1969 (shown as the dashed line). The colonies grown at 24°C died in July 1970; daily records made at 24°C were continued again in October 1970 after other colonies were transferred from 17 to 24°C. Yearly changes in the coastal water temperature at Woods Hole, Mass. were calculated from daily records compiled from 1967 through 1969 by the U.S. Coast and Geodetic Survey. (From Brock, M. A., *Comp. Biochem. Physiol.*, 51A, 391, 1975. With permission.)

## VI. DISCUSSION

Throughout this presentation, senescence and its associated changes in several parameters have been described for adult hydranths of colonial cnidarians, with the assumption that the hydranths may be considered individual organisms. Whether to accept this view or the contention that the colony should be considered an individual organism is a dilemma. Hyman[62] stated that "Hydroid colonies are polymorphic, i.e., consist of more than one kind of individual (called *zooid* or *person*)." Crowell[16] wrote "a hydroid colony is derived from a zygote or from a fragment of another colony and hence may be regarded as an individual possessing many duplicated organs, the polyps or hydranths. Zoological tradition, however, considers a polyp to be an individual and the assemblage of such polyps a colony." The bias of many who have observed cnidarian colonies for several years is to consider the adult hydranths as individuals, and reasons bearing on this will be given later. The coelenteron or digestive cavity of each hydranth, however, is continuous with that of the whole colony, and time lapse photography of Hammett[12] showed "hydroplasmic streaming" which carried particles throughout the colony, primarily toward developing hydranth buds and both toward and away from adult hydranths. Regression fragments leave the regressing hydranths and are phagocytized by the digestive cells of adults in the colony.[18,19] Experimental evidence that the nutritional level of the colony, that is the feeding of adult hydranths, influences cellular proliferation at distant growth zones was provided by Crowell.[16] Therefore, the regulation of growth and the development of hydranth buds are related to and depend on colonial function.

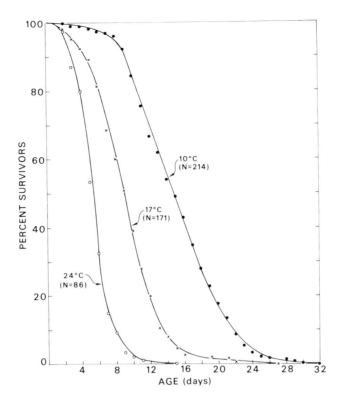

FIGURE 18. The effect of three different environmental temperatures on the longevity of *C. flexuosa* hydranths during November 1970 is shown for colonies cultured in 10, 17, and 24°C constant environments. (From Brock, M. A., *Comp. Biochem. Physiol.*, 51A, 391, 1975. With permission.)

The earliest stage of development in which the individuality of the hydranths may be suspected is just before the fully differentiated tentacles extend. One such organism may become a feeding adult while its nearest neighbor in the same developmental stage is autolyzed.[17] Mature hydranths surely seem to function as individual organisms. Adults in adjacent locations may exhibit vastly different lifespans, the rate of senecence of one not affecting the other. They were isolated from the colony and regressed normally several days later; all produced normal stolons at the cut end and 25% produced hydranths.[63] More convincing evidence may be in the analysis of the electrical potentials correlated with behavioral responses of *Obelia geniculata* hydranths.[49] Recordings of these potentials, thought to be epithelial potentials as previously mentioned, showed that there was no coupling of contraction potential activity between hydranths without applied stimuli. The authors[49] summarized by stating '' . . . these potentials are not normally through-conducted between hydranths, but are instead local events within a hydranth. Thus each hydranth acts as an individual unit rather than part of a tightly coupled colony.'' Through-conduction in an *O. geniculata* colony was observed only after electrical or mechanical stimulation of a hydranth.[50] Similarly, the stimulation of one hydranth of a *Clava squamata* colony was followed by the excitation of a neighboring hydranth,[64] and it was suggested that the stimulus-induced, coordinated responses of the hydranths provide a protective advantage for the colony. The unresolved dilemma is further embellished with the knowledge that both adult hydranths and entire colonies of *C. flexuosa* have finite lifespans.

Although more complex nervous systems including endogenous neural pacemakers and sensory receptors exist in some cnidarians,[65] the persistent circannual rhythmicity was ob-

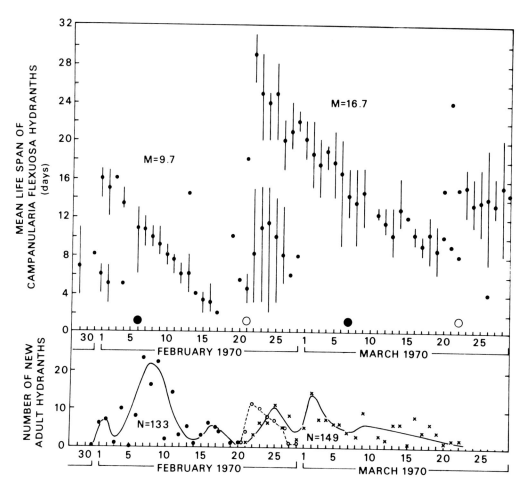

FIGURE 19.    The mean and range of lifespans for adult hydranths of *C. flexuosa* that matured on each successive day (closed circles) are shown in the upper section of the figure. The large solid and open circles represent nights of new moon and full moon, respectively. The number of hydranths that became adults on each successive day is shown in the lower section of the figure. Hydranths that matured between February 1 and 17 (closed circles) had a mean lifespan of 9.7 days; adults that matured between February 22 and March 22 (crosses) had a mean lifespan of 16.7 days. Some hydranths that matured between February 20 and 28 lived about half as long as the longest-lived adults. Their lifespans are shown in the upper section of the figure, and the number of hydranths that matured in that group is indicated by the open circles connected with a broken line. (From Brock, M. A., *J. Interdiscipl. Cycle Res.*, 7, 269, 1976. With permission.)

served in a hydrozoan with a deceivingly simple nerve net. Endogenous rhythms with such long periodicities have been described only in more highly evolved organisms such as crayfish, birds, and mammals,[66-68] and the "lifespan clock" hypothesized by Comfort,[3] which depends on a neuroendocrine system, was most plausibly localized to the hypothalamus. This degree of specialization is obviously unnecessary for the expression of circasemilunar and circannual rhythmicity in *C. flexuosa*. Comfort's suggestion that the lifespan is limited by the "general interplay of homeostatic failures" and his discussion of "brain loci acting as pacemakers of aging" might be restated using different terms. In a 1960 discussion of the circadian organization of living systems, Pittendrigh[69] proposed that the "loss of proper phasing among physiological subsystems is detrimental," and he, Aschoff, and associates were later able to show experimentally that disruption of the normal circadian organization of *Drosophila melanogaster* and *Phormia terraenovae* was associated with reduced longevity of the flies.[70-72] Internal desynchronization of circadian rhythms was

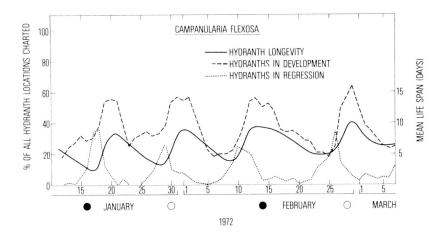

FIGURE 20. *C. flexuosa* colonies obtained from Woods Hole, Mass. in late November 1971 were grown in a constant 17°C, DD, environment. Each day, 34 to 204 locations were charted, with the smallest number observed just after the peaks in regression. The large solid and open circles represent nights of new moon and full moon, respectively.

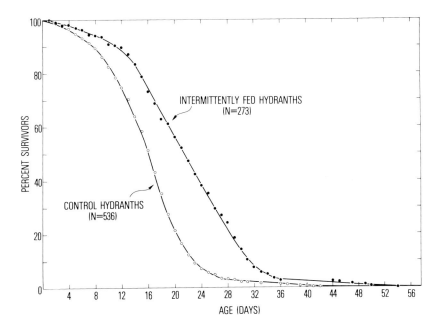

FIGURE 21. Survivorship of *C. flexuosa* hydranths fed intermittently during October and November 1969 as compared with hydranths fed daily at 10°C. The mean lifespan of 16.4 days for the control animals was increased to 21.0 days in the group fed every 3 days. The average number of brine shrimp ingested by individual hydranths during one feeding period was 13.4 for the controls and 14.3 for the hydranths fed intermittently. (N = 60 for both groups.)

initiated either by subjecting the flies to repeated phase shifts simulating trans-Atlantic air trips or to light-dark cycles whose periods were longer or shorter than an optimal 24-hr period. Their views that maintaining proper phase relationships among circadian rhythms results in normal physiological function also are applicable to cnidarians, because it was observed that desynchronization of circasemilunar rhythmicity in regression and development of *C. flexuosa* hydranths occurred just before the death of the colonies grown at 17°C.[26]

Perhaps the most interesting aspect of senescence in *C. flexuosa* and possibly other cnidarians is that the rate of aging is modified both by extrinsic factors such as ambient temperature and intermittent feeding and by endogenous rhythms. The rhythmicity in longevity, which is probably an evolutionary adaptation to lunar and annual periodicities in the environment, provides a model in which natural changes in the rates of development and aging may be studied. Since the various cellular alterations associated with age summarized in this paper were documented for only one phase of the circannual rhythm, the question of whether differences in rates and/or patterns of aging exist in other phases remains unanswered.

# ACKNOWLEDGMENTS

The assistance of Malcolm Gee, Gilbert Press, and Bruce Troup in some of these studies is gratefully acknowledged. The drawings and much of the photography were done by Charlotte Adler, William Fisher, Kevin Lewis, and Rowland Schnick.

# REFERENCES

1. **Crowell, S.,** The regression-replacement cycle of hydranths of *Obelia* and *Campanularia*, *Physiol. Zool.*, 26, 319, 1953.
2. **Brock, M. A. and Strehler, B. L.,** Studies on the comparative physiology of aging. IV. Age and mortality of some marine cnidaria in the laboratory, *J. Gerontol.*, 18, 23, 1963.
3. **Comfort, A.,** *The Biology of Senescence*, Elsevier/North-Holland, N.Y., 1979, chap. 2,5,6,9.
4. **Strehler, B. L.,** *Time, Cells and Aging*, Academic Press, N.Y., 1977, chap. 3,4.
5. **Ashworth, J. H. and Annandale, N.,** Observations on some aged specimens of *Sagartia troglodytes*, and on the duration of life in coelenterates, *Proc. R. Soc. Edinburgh*, 25, 295, 1904.
6. **Minasian, L. L., Jr.,** The distribution of proliferating cells in an anthozoan polyp, *Haliplanella luciae* (Actiniaria: Acontiaria), as indicated by $^3$H-thymidine incorporation, in *Developmental and Cellular Biology of Coelenterates*, Tardent, P. and Tardent, R., Eds., Elsevier/North-Holland, Amsterdam, 1980, 415.
7. **Brien, P.,** La pérennité somatique, *Biol. Rev.*, 28, 308, 1953.
8. **Shostak, S., Patel, N. G., and Burnett, A. L.,** The role of mesoglea in mass cell movement in *Hydra*, *Dev. Biol.*, 12, 434, 1965.
9. **Campbell, R. D.,** Tissue dynamics of steady state growth in *Hydra littoralis*. II. Patterns of tissue movement, *J. Morphol.*, 121, 19, 1967.
10. **Campbell, R. D.,** Vital marking of single cells in developing tissues: India ink injection to trace tissue movements in *Hydra*, *J.Cell. Sci.*, 13, 651, 1973.
11. **Campbell, R. D.,** Tissue dynamics of steady state growth in *Hydra littoralis*. I. Patterns of cell division, *Dev. Biol.*, 15, 487, 1967.
12. **Hammett, F. S.,** The role of the amino acids and nucleic acid components in developmental growth. I. The growth of an *Obelia* hydranth, *Growth*, 7, 331, 1943.
13. **Berrill, N. J.,** The polymorphic transformations of *Obelia*, *Q. J. Microsc. Sci.*, 90, 235, 1949.
14. **Berrill, N. J.,** Growth and form in calyptoblastic hydroids. I. Comparison of a campanulid, campanularian, sertularian, and plumularian, *J. Morphol.*, 85, 297, 1949.
15. **Berrill, N. J.,** Growth and form in calyptoblastic hydroids. II. Polymorphism within the *Campanularidae*, *J. Morphol.*, 87, 1, 1950.

16. **Crowell, S.,** Differential responses of growth zones to nutritive level, age, and temperature in the colonial hydroid *Campanularia, J. Exp. Zool.,* 134, 63, 1957.

17. **Brock, M. A.,** Circannual rhythms. I. Free-running rhythms in growth and development of the marine cnidarian, *Campanularia flexuosa, Comp. Biochem. Physiol.,* 51A, 377, 1975.

18. **Huxley, J. S. and deBeer, G. R.,** Studies in dedifferentiation. IV. Resorption and differential inhibition in *Obelia* and *Campanularia, Q. J. Microsc. Sci.,* 67, 473, 1923.

19. **Lunger, P. D.,** Fine-structural aspects of digestion in a colonial hydroid, *J. Ultrastruct. Res.,* 9, 362, 1963.

20. **Brock, M. A., Strehler, B. L., and Brandes, D.,** Ultrastructural studies on the life cycle of a short-lived metazoan, *Campanularia flexuosa.* I. Structure of the young adult, *J. Ultrastruct. Res.,* 21, 281, 1968.

21. **Brock, M. A.,** Ultrastructural studies on the life cycle of a short-lived metazoan, *Campanularia flexuosa.* II. Structure of the old adult, *J. Ultrastruct. Res.,* 32, 118, 1970.

22. **Chapman, D. M.,** Cnidarian histology, in *Coelenterate Biology — Reviews and New Perspectives,* Muscatine, L. and Lenhoff, H. M., Eds., Academic Press, N.Y., 1974, chap. 1.

23. **Sabatini, D. D., Bensch, K., and Barrnett, R. J.,** Cytochemistry and electron microscopy — the preservation of cellular ultrastructure and enzymatic activity by aldehyde fixation, *J. Cell Biol.,* 17, 19, 1963.

24. **Millonig, G.,** Further observations on a phosphate buffer for osmium solutions, *5th Int. Congr. Electron Microscopy,* P-8, 1962.

25. **Caulfield, J. B.,** Effects of varying the vehicle for $OsO_4$ in tissue fixation, *J. Biophys. Biochem. Cytol.,* 3, 827, 1957.

26. **Brock, M. A.,** Unpublished data, 1983.

27. **Holtzman, E.,** *Lysosomes: A survey,* Springer-Verlag, N.Y., 1976, chap. 3.

28. **Farquhar, M. G.,** Lysosome function in regulating secretion: disposal of secretory granules in cells of the anterior pituitary gland, in *Lysosomes in Biology and Pathology,* Vol. 2, Dingle, J. T. and Fell, H. B., Eds., Elsevier/North-Holland, Amsterdam, 1969, chap. 16.

29. **West, D. L.,** The epitheliomuscular cell of *Hydra:* its fine structure, three-dimensional architecture, and relation to morphogenesis, *Tissue Cell,* 10, 629, 1978.

30. **Filshie, B. K. and Flower, N. E.,** Junctional structures in *Hydra, J. Cell Sci.,* 23, 151, 1977.

31. **Tidbull, J. G.,** An ultrastructural and cytochemical analysis of the cellular basis for tyrosine-derived collagen crosslinks in *Leptogorgia virgulata* (Cnidaria: Gorgonacea), *Cell Tissue Res.,* 222, 635, 1982.

32. **Mariscal, R. N.,** The elemental composition of nematocysts as determined by X-ray microanalysis, in *Developmental and Cellular Biology of Coelenterates,* Tardent, P. and Tardent, R., Eds., Elsevier/North-Holland, Amsterdam, 1980, 337.

33. **Pearse, A. G. E.,** *Histochemistry Theoretical and Applied,* Vol. 1, 3rd ed., Williams & Wilkins, Baltimore, 1968, 556, 601, 728.

34. **Chen, P. S., Toribara, T. Y., and Warner, H.,** Microdetermination of phosphorus, *Anal. Chem.,* 28, 1756, 1956.

35. **Kissane, J. M. and Robins, E.,** The fluorometric measurement of deoxyribonucleic acid in animal tissues with special reference to the central nervous system, *J. Biol. Chem.,* 233, 184, 1958.

36. **Campbell, R. D.,** Cell proliferation and morphological patterns in the hydroids *Tubularia and Hydractinia, J. Embryol. Exp. Morphol.,* 17, 607, 1967.

37. **Östman, C.,** Isoenzymes and taxonomy in Scandinavian hydroids (Cnidaria, Campanulariidae), *Zool. Scr.,* 11, 155, 1982.

38. **Strehler, B. L. and Crowell, S.,** Studies on comparative physiology of aging. I. Function vs. age of *Campanularia flexuosa, Gerontologia,* 5, 1, 1961.

39. **Lowry, O. H., Rosebrough, N. J., Farr, A. L., and Randall, R. J.,** Protein measurement with the Folin phenol reagent, *J. Biol. Chem.,* 193, 265, 1951.

40. **David, C. N.,** A quantitative method for maceration of *Hydra* tissue, *Wilhelm Roux Arch. Entwicklungsmech. Org.,* 171, 259, 1973.

41. **Bode, H., Berking, S., David, C. N., Gierer, A., Schaller, H., and Trenkner, E.,** Quantitative analysis of cell types during growth and morphogenesis in *Hydra, Wilhelm Roux Arch. Entwicklungsmech. Org.,* 171, 269, 1973.

42. **David, C. N. and Campbell, R. D.,** Cell cycle kinetics and development of *Hydra attenuata.* I. Epithelial cells, *J. Cell Sci.,* 11, 557, 1972.

43. **Campbell, R. D. and David, C. N.,** Cell cycle kinetics and development of *Hydra attenuata.* II. Interstitial cells, *J. Cell Sci.,* 16, 349, 1974.

44. **David, C. N. and Gierer, A.,** Cell cycle kinetics and development of *Hydra attenuata.* III. Nerve and nematocyte differentiation, *J. Cell Sci.,* 16, 359, 1974.

45. **Sproull, F. and David, C. N.,** Stem cell growth and differentiation in *Hydra attenuata.* II. Regulation of nerve and nematocyte differentiation in multiclone aggregates, *J. Cell Sci.,* 38, 171, 1979.

46. **McNeil, P. L.,** Mechanisms of nutritive endocytosis. I. Phagocytic versatility and cellular recognition in *Chlorohydra* digestive cells, a scanning electron microscope study, *J. Cell Sci.,* 49, 311, 1981.

47. **Bode, H. R., Flick, K. H., and Bode, P. M.,** Constraints on the relative sizes of the cell populations in *Hydra attenuata, J. Cell Sci.,* 24, 31, 1977.

48. **Mariscal, R. N.,** Nematocysts, in *Coelenterate Biology — Reviews and New Perspectives,* Muscatine, L. and Lenhoff, H. M., Eds., Academic Press, N.Y., 1974, chap. 3.

49. **Morin, J. G. and Cooke, I. M.,** Behavioral physiology of the colonial hydroid *Obelia.* I. Spontaneous movements and correlated electrical activity, *J. Exp. Biol.,* 54, 689, 1971.

50. **Morin, J. G. and Cooke, I. M.,** Behavioral physiology of the colonial hydroid *Obelia.* II. Stimulus-initiated electrical activity and bioluminescence, *J. Exp. Biol.,* 54, 707, 1971.

51. **Morin, J. G. and Cooke, I. M.,** Behavioral physiology of the colonial hydroid *Obelia.* III. Characteristics of the bioluminescent system, *J. Exp. Biol.,* 54, 723, 1971.

52. **Brock, M. A.,** Circannual rhythms. III. Temperature-compensated free-running rhythms in growth and development of the marine cnidarian, *Campanularia flexuosa, Comp. Biochem. Physiol.,* 51A, 385, 1975.

53. **Brock, M. A.,** Circannual rhythms. II. Rhythmicity in the longevity of hydranths of the marine cnidarian, *Campanularia flexuosa, Comp. Biochem. Physiol.,* 51A, 391, 1975.

54. **Brock, M. A.,** Free-running rhythmicity in the life spans of hydranths of the marine cnidarian, *Campanularia flexuosa, J. Interdiscip. Cycle Res.,* 7, 269, 1976.

55. **Brock, M. A.,** Differential sensitivity to temperature steps in the circannual rhythm of hydranth longevity in the marine cnidarian, *Campanularia flexuosa, Comp. Biochem. Physiol.,* 64A, 381, 1979.

56. **Brock, M. A.,** Comparison of circasemilunar rhythmicity in east and west coast cnidarians, in *Developmental and Cellular Biology of Coelenterates,* Tardent, P. and Tardent, R., Eds., Elsevier/North-Holland, Amsterdam, 1980, 23.

57. **Aschoff, J., Klotter, K., and Wever, R.,** Circadian vocabulary, in *Circadian Clocks,* Aschoff, J., Ed., Elsevier/North-Holland, Amsterdam, 1965, p. x (preface).

58. **Pittendrigh, C. S.,** On temperature independence in the clock system controlling emergence time in *Drosophila, Proc. Natl. Acad. Sci. U.S.A.,* 40, 1018, 1954.

59. **Sweeney, B. M. and Hastings, J. W.,** Effects of temperature upon diurnal rhythms, *Cold Spring Harbor Symp. Quant. Biol.,* 25, 87, 1960.

60. **Aschoff, J.,** Adaptive cycles: their significance for defining environmental hazards, *Int. J. Biometeor.,* 11, 255, 1967.

61. **Hoffmann, K.,** The adaptive significance of biological rhythms corresponding to geophysical cycles, in *The Molecular Basis of Circadian Rhythms,* Hastings, J. W. and Schweiger, H.-G., Eds., Dahlem Konferenzen Berlin, 1976, 63.

62. **Hyman, L. H.,** *The Invertebrates: Protozoa through Ctenophora,* McGraw-Hill, N.Y., 1940, 407.

63. **Nathanson, D. L.,** The relationship of regenerative ability to the regression of hydranths of *Campanularia, Biol. Bull.,* 109, 350, 1955.

64. **Stokes, D. R. and Rushforth, N. B.,** Evoked responses to electrical stimulation in the colonial hydroid *Clava squamata:* a contraction pulse system, *Biol. Bull.,* 157, 189, 1979.

65. **Josephson, R. K.,** Cnidarian Neurobiology, in *Coelenterate Biology — Reviews and New Perspectives,* Muscatine, L. and Lenhoff, H. M., Eds., Academic Press, N.Y., 1974, chap. 6.

66. **Jegla, T. C. and Poulson, T. L.,** Circannian rhythms. I. Reproduction in the cave crayfish, *Orconectes pellucidus inermis, Comp. Biochem. Physiol.,* 33, 347, 1970.

67. **Berthold, P.,** Circannual rhythms in birds with different migratory habits, in *Circannual Clocks — Annual Biological Rhythms,* Pengelley, E. T., Ed., Academic Press, N.Y., 1974, 55.

68. **Pengelley, E. T. and Asmundson, S. J.,** Circannual rhythmicity in hibernating mammals, in *Circannual Clocks — Annual Biological Rhythms,* Pengelley, E. T., Ed., Academic Press, N.Y., 1974, 95.

69. **Pittendrigh, C. S.,** Circadian rhythms and the circadian organization of living systems, *Cold Spring Harbor Symp. Quant. Biol.,* 25, 159, 1960.

70. **Aschoff, J., Saint Paul, U. v., and Wever, R.,** Die Lebensdauer von Fliegen unter dem Einfluss von Zeit-Verschiebungen, *Naturwissenschaften,* 58, 574, 1971.

71. **Pittendrigh, C. S. and Minis, D. H.,** Circadian systems: longevity as a function of circadian resonance in *Drosophila melanogaster, Proc. Natl. Acad. Sci. U.S.A.,* 69, 1537, 1972.

72. **Saint Paul, U. v. and Aschoff, J.,** Longevity among blowflies *Phormia terraenovae R.D.* kept in non-24-hour light-dark cycles, *J. Comp. Physiol.,* 127, 191, 1978.

Chapter 3

# CELL-SPECIFIC GENE EXPRESSION IN *CAENORHABDITIS ELEGANS*

**Michael R. Klass**

## TABLE OF CONTENTS

## I. INTRODUCTION AND DESCRIPTION OF THE ORGANISM

An ideal experimental system for the study of aging and development is one that is accessible by all levels of analysis — genetic, cellular, biochemical, and molecular — while at the same time exhibiting the major aspects of development and aging common to most metazoans. The free-living nematode *Caenorhabditis elegans* offers the experimenter an excellent opportunity for such a multidisciplinary approach.

*C. elegans* is a multicellular eucaryote that differentiates specific tissue types; hypodermal, muscle, nerve, intestinal, and gonadal tissues are all present in the worm. Therefore, by implication, it undergoes extensive and specific gene regulation.

The worm can be easily cultured in the laboratory either axenically or monoxenically and large quantities can be obtained for biochemical analysis. It has a transparent body with a simple anatomy that has been characterized in detail. All of the somatic cell lineages and cell migrations have been recorded. Of greatest importance from a genetic standpoint is that *C. elegans* is an hermaphroditic, genetically manipulable organism that facilitates the combination of biochemical and mutational analysis.

### A. Laboratory Culture Methods

*C. elegans* can be easily cultured in the laboratory and is routinely grown on agar plates and fed *E. coli*.[1] Optimum growth temperature is 20°C. Permissive and restrictive temperatures used for temperature-sensitive mutants are 16 and 25.5°C, respectively. The worm can also be grown either in a rich chemical medium[2-4] or in liquid medium supplemented with *E. coli* and large quantities can be obtained for biochemical studies (6 g wet weight per liter). All wild-type and mutant strains can be stored indefinitely in liquid nitrogen and maintain a 10 to 50% viability.

### B. Anatomy and Development

One of the major advantages of the nematode as an experimental system is its transparent cuticle and simple anatomy. This advantage has facilitated the complete recording of the entire cell lineage from fertilization to the adult.[5-8] The eggs of *C. elegans* are prolate spheroids approximately $40 \times 60$ $\mu$m surrounded by a transparent shell. After embryogenesis, a first stage larva (L1) hatches from the egg. An L1 contains 558 somatic cells and a four-cell gonadal primordium.[5] Somatic cells consist of nervous, gonadal, intestinal, and hypodermal tissues. The first stage larva (240 $\mu$m long) begins to feed and develop, undergoing four larval molts as it grows to a sexually mature adult. Growth and development to maturation takes place in approximately 72 hr at 20°C. If, during its early development, the larva is subjected to starvation conditions it will enter a developmentally arrested state, called the dauer larva stage, during the L2 to L3 molt. Dauer larvae live for extended periods of time, an average of 45 days at 20°C, compared to 14 days for nondauer worms. Upon refeeding, dauer larvae enter the normal developmental pathway and grow to be sexually mature adults[9] and display normal adult lifespans.[10]

A sexually mature adult hermaphrodite is approximately 1200 $\mu$m in length and 100 $\mu$m in diameter (Figure 1). The adult hermaphrodite contains 953 somatic cells and a gonad containing approximately 2600 nuclei.[5,11] The worm is surrounded by a collagenous cuticle covered by a proteinaceous cortical layer.[12,13] The nervous system[14-18] and the body musculature[19-24] have been studied in great detail.

As a sexually mature adult, *C. elegans* produces approximately 300 progeny by self-fertilization over a $3\frac{1}{2}$ day period. The mean lifespan of hermaphrodites is 23 days at 16°C, 14.5 days at 20°C, and 8.9 days at 25.5°C, with some variation encountered between experiments.[25] The hermaphrodite reproductive system is composed of a bilateral symmetrical structure that produces sperm during the 3rd and 4th larval stages and then oocytes, which

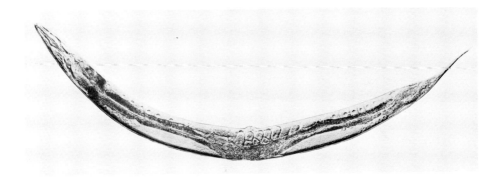

A

LOOP

PROXIMAL ARM

Vulva
Vagina
Zygotes
Uterus
Spermatheca
Oocytes

DISTAL ARM

Ovary

100μ

B

FIGURE 1.   (A) This is a differential interference contrast photograph of a *C. elegans* adult hermaphrodite.
(B) This is a diagrammatic representation of A showing the details of the hermaphrodite gonad.

continue to be produced from late in the 4th larval stage until after sperm are exhausted.[11,26-27] Mature sperm are housed in the spermathecae and used to fertilize oocytes as they pass through to the uterus. The two ovaries are located on the dorsal side of the worm and each is connected to a spermatheca by a reflexed oviduct. Maturing oocytes pass down the oviduct to the spermatheca, are fertilized and enter the uterus where they undergo their first embryonic cleavages, and finally exit out the vulva located midway on the ventral side of the worm. The exact cell lineage pattern for all the somatic components of the reproductive system in both the hermaphrodite and the male is now known.[6]

Males arise by a process of nondisjunction of the X chromosomes during meiosis in the hermaphrodite. Hermaphrodites are XX and males are XO. Immediately after hatching, the male nematode contains a four-cell gonadal primordium indistinguishable from that found in the hermaphrodite. However, in the male, this primordium develops into an asymmetrical gonad with only a single testis and one reflexed arm (Figure 2). The process of spermatogenesis begins in the testis region and proceeds as developing spermatocytes pass through the loop to the seminal vesicle where mature sperm are housed. During mating, mature sperm are passed from the seminal vesicle through the *vas deferens* and injected into the hermaphrodite uterus. The male gonad continues the process of spermatogenesis until at least 10 days after hatching at 20°C.

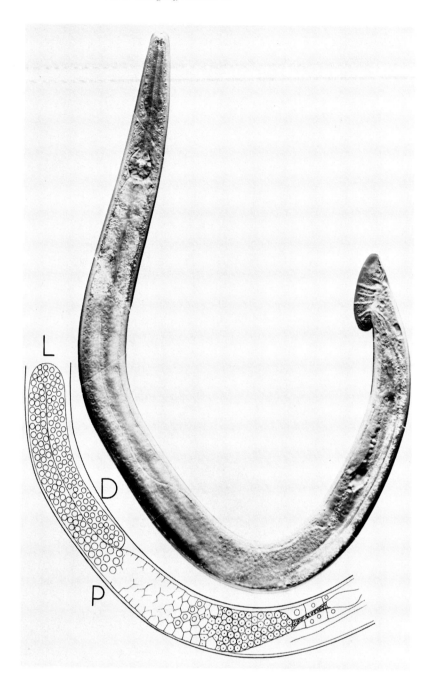

FIGURE 2.    This is a photograph of an adult male of *C. elegans* taken with differential interference optics. Also shown is a diagram of the gonad. (D) Distal tip of gonad; (L) loop region; (P) proximal arm.

## C. Genetics

*C. elegans* is usually a diploid hermaphroditic organism that possesses 5 pairs of autosomes and a pair of X (sex) chromosomes. The usual mode of reproduction is by self-fertilization of the hermaphrodite.[29] A newly induced mutation can be made homozygous by self-fertilization. The F1 progeny of the mutated parental generation are heterozygous for the mutation

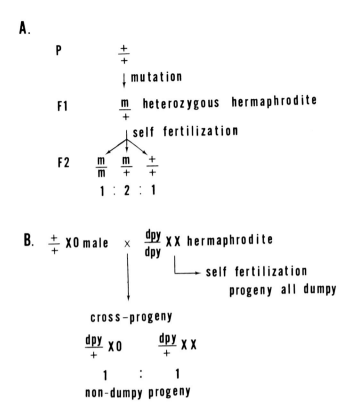

FIGURE 3. (A) Diagram of the hermaphroditic mode of inheritance in
*C. elegans*. (M) putative recessive mutation; (P) parental generation; (F1)
first generation; (F2) second generation. (B) Diagram of mating between
wild-type male and a homozygpous recessive dpy/dpy hermaphrodite; dpy,
recessive mutation causing "dumpy" phenotype.

and will produce one fourth homozygous mutant, one half heterozygous mutant, and one fourth homozygous wild-type progeny in the F2 generation (see Figure 3A).

Crosses can be performed with male nematodes. Males arise spontaneously by a process of nondisjunction of the X chromosomes during meiosis in the hermaphrodite.[30,31] Males are XO and are easily recognized by the copulatory bursa on the tail (Figure 2). In such crosses both self-progeny and cross-progeny may be produced, so hermaphrodites homozygous for a recessive-maker are used to distinguish cross-progeny from self-progeny. Alternatively, genetically self-sterile hermaphrodites can be used to produce 100% cross-progeny. Of the cross-progeny, 50% are XO males which are used in crosses for routine genetic manipulations. This simple form of inheritance, together with a short generation time of 3 days at 20°C, makes *C. elegans* ideally suited for genetic studies.

Brenner[1] initiated genetic analysis of *C. elegans* by characterizing about 300 mutations defining about 100 genes which were induced by the mutagen ethylmethanesulfonate (EMS). Brenner[1] calculated an average forward mutation rate of $5 \times 10^{-4}$ per gene by dividing the frequency of appearance of mutations by the total number of complementation groups. The ability for detailed genetic analysis is now being extended by the isolation of balancer stocks, deletions, duplications, and lethal mutations,[32,33] as well as the recent identification of the only mutation identified in a metazoan which causes the synthesis of an amber suppressor tRNA.[34,35] The genetics of *C. elegans* has been reviewed by Herman and Horvitz[36] and the genetic analysis of aging is described by Johnson in Chapter 4 of this book.

All of the mutant strains that have been characterized and mapped are available from the *Caenorhabditis elegans* Stock Center at the University of Missouri at Columbia. Hundreds of mutants have been isolated and characterized that affect various aspects of gonadogenesis, spermatogenesis, development, chemotaxis, dauer larva formation, sexual dimorphism, cell cycle, and the nervous system.[11,31,37-42] Approximately 300 temperature-sensitive *(ts)* mutants that interrupt the life cycle of *C. elegans* have been isolated and used to analyze development by Hirsh and Vanderslice;[11,37,43] 25 of these *ts* mutants have been identified as spermatogenesis-defective by the following criteria. These mutants fail to reproduce at restrictive temperature and lay unfertilized oocytes; however, they can reproduce if mated with wild-type males. Use of these and other mutants to help study cellular differentiation in *C. elegans* is described in the next section.

## II. *C. ELEGANS* AS AN EXPERIMENTAL SYSTEM FOR CELL-SPECIFIC GENE EXPRESSION

These characteristics of *C. elegans* and the large amount of data gathered on its development and genetics make it an excellent system for the genetic and biochemical analysis of cellular differentiation.[44] One approach to understanding cell differentiation is to analyze the mechanisms involved in cell-specific gene expression. More specifically, we wish to know why certain genes are active in only specific cell types and not in others. The first step in such an approach is the genetic and biochemical analysis of differentiated cell types and their specific gene products. Current studies on oocytes, sperm, hypodermal, muscle, and intestinal cells of *C. elegans* are described below.

### A. Oocytes

Oocytes in the nematode have been shown to contain at least four specific proteins that are synthesized in abundance at the time of oogenesis.[46-48] In mutant strains displaying abnormal sexual development, the appearance of these proteins is correlated with the formation of oocytes. For example, mutations in any of the *tra* genes (*tra-1*, *tra-2*, or *tra-3*) cause genotypically XX hermaphrodites to develop as phenotypic males.[27,30,31] All of the mutant alleles at the *tra-2* locus cause the transformation of genotypic hermaphrodites into adult worms with varying degrees of maleness while some of the mutant alleles at the *tra-1* locus cause transformation into phenotypically viable, fertile males. One of these mutant strains (DH202) demonstrates a temperature-sensitive phenotype that causes transformation only at the restrictive temperature.[27] Using specific periods of restrictive temperature it is possible to cause the development of intersexes showing varying degrees of maleness. In the predominantly male intersex, the adult has a male gonad that produces oocytes. The four oocyte-specific proteins are present in this male intersex. In the predominantly hermaphrodite intersex, however, the adult has a hermaphrodite gonad that produces only sperm. The four oocyte-specific proteins are absent in this hermaphrodite intersex.[46]

Isolation of oocytes is necessary for biochemical analysis. It is difficult to obtain oocytes from the oviduct of the hermaphrodite except by dissection of individual worms. However, using a spermatogenesis-defective mutant strain (DH26), unfertilized oocytes that have entered the uterus can be obtained. Due to the lack of viable sperm in this mutant, oocytes will pass through the spermatheca and into the uterus without being fertilized. Large synchronous cultures of the temperature-sensitive spermatogenesis mutant are grown at the restrictive temperature during the temperature-sensitive period for spermatogenesis, causing the production of defective sperm. The worms are then collected by washing on a 20-$\mu$m nylon Nitex filter (Tetko, Inc. Elmsford, N.Y.) which allows the smaller bacteria and debris to wash through while retaining the worms. The oocytes are removed by squashing the worms between two plexiglass (6 × 10 in.) plates in 5 m$M$ Hepes pH 7.2, 110 m$M$ NaCl,

4 m$M$ KCl, 5 m$M$ Mg$^{++}$, and 5 m$M$ Ca$^{++}$ at 1000 to 2000 psi. The plexiglass plates are then separated and rinsed off with the same buffer. The rinse is filtered through 20-μm nylon filters. The released oocytes pass through the filter while the worm carcasses do not. The oocytes then are pelleted at low speed in a clinical centrifuge, washed, and the pellets pooled on ice. The oocytes are then allowed to settle, producing pellets of uncontaminated oocytes. These oocytes are currently being used to assay maternal enzyme activities.

One problem encountered with oocyte preparation is that many of the oocytes, once they have passed the spermatheca, appear to undergo chromosomal endoreduplication and contain enlarged nuclei and excessive amounts of yolk and rough ER.[46] This problem is somewhat alleviated by utilizing a double mutant (CB146, DH232) that is spermatogenesis-defective but also produces oocytes that do not undergo endoreduplication.

There are four oocyte-specific proteins that have been isolated and shown to be yolk proteins. Isolated yolk proteins from whole worms have been used to obtain specific antibody. Sharrock and Kimble[47,48] have shown that these yolk proteins are synthesized in the intestine and transferred to the gonad and eventually to the oocytes. This observation has led to the suggestion that the intestinal cells may play a role similar to nurse cells in insect oogenesis. Five yolk protein genes have been cloned and are currently being used to analyze the tissue-specific expression of this gene family.[73]

## B. Sperm

Mature sperm of most nematode species lack flagella and move by amoeboid motion.[49,50] Pseudopodial movement of *C. elegans* sperm has been described in detail.[51-54] Sperm cells of the nematode *C. elegans* can be obtained in sufficient quantities for biochemical analysis and have provided the basis for the study of the regulation of sperm-specific proteins.[55] Both spermatids and mature sperm of *C. elegans* contain a number of characteristic subcellular organelles. The spermatid has a condensed nucleus with clearly identifiable centrioles that can also be observed in mature sperm. Mitochondria are visible in the spermatid but the cytoplasm consists mainly of fibrous bodies with their attached vesicles. During spermatogenesis, the membranes around the fibrous bodies disappear while there is an increase in the amount of invaginated membranes associated with the vesicles.

The mature sperm has dense cytoplasm, mitochondria, and special membrane structures resembling acrosomes located near the periphery of the cell. The mature sperm nucleus is condensed and does not have a distinct nuclear envelope. An opaque halo also surrounds the nucleus, and the centrioles are located within this halo. The entire process of spermatogenesis has been studied in detail by light microscopy[27,55] and at the electron microscope level in mature males.[28]

Biochemical quantities of sperm cells can be isolated from the nematode through the use of *him* mutants that cause large increases in the rate of meiotic nondisjunction of the X chromosomes in the hermaphrodite.[30] These *him* mutants produce up to 36% male (XO) progeny in each generation. Mass cultures of *him* mutants are grown on agar plates supplemented with 6 mℓ of a mixture of 1 chicken egg and 50 mℓ of nutrient broth and seeded with *E. coli*. Adult worms are gently washed off the plates, collected on 20-μm nylon filter screens and washed extensively with M9 salt solution[1] to remove excess bacteria. Washed worms are then layered on a 35-μm nylon filter screen. The males (mean diameter ≤40 μm) crawl through the filter while the hermaphrodites (mean diameter ≥40 μm) do not. Routinely, cultures of 1 to 5 × 10$^6$ worms with 70 to 80% males can be obtained in this manner. This enriched male culture is then washed again and used as a source of sperm for the following procedure.

Worms from an enriched male culture are squashed between two plexiglass plates of equal size in a Carver Laboratory Press at 10,000 psi. This pressure forces the males to release their sperm through the cloaca at the tail. The plates are then rinsed off with cold M9 salt solution and the rinse, containing worm carcasses and released sperm, is filtered through 10-μm nylon filters. The sperm (diameter 5 to 6 μm) pass through but the worm carcasses and large debris are collected on the filter. Sperm are then pelleted in a clinical centrifuge at low speed, washed in M9 salts, and filtered through an 8-μm polycarbonate Nucleopore filter to remove larger debris. This method routinely produces $1 \times 10^8$ sperm.

During development of the male nematode, a basic low molecular weight protein, whose appearance corresponds to the initial stages of spermatogenesis, is synthesized. Biochemical fractionation of purified sperm shows that this low molecular weight protein is a major constituent of sperm, accounting for approximately 17% of the total sperm protein. This protein appears to be a homodimer with a 15,000 monomer molecular weight. This major sperm protein (called MSP) is a basic protein with an isolectric point of 8.6. It is a very soluble sperm component that can be easily purified by gentle homogenization in a low salt buffer, centrifugation at 100,000 g, followed by precipitation of acidic protein at pH 4.3 and binding to phosphocellulose.[55]

Immunocytochemical localization using antibody specific to MSP has shown this protein to be a specific product of spermatogenesis. It can be detected only in the proximal arm of the male gonad and in mature sperm in both the male and hermaphrodite. Using a technique described by Mackenzie et al.,[56] adult males and hermaphrodites were fixed in 1.5% paraformaldehyde and embedded in polyethylene glycol 4000 for analysis by immunocytochemical techniques. Figure 4 shows the localization of MSP in the spermatheca of the adult hermaphrodite by the immunoperoxidase technique. Only sperm and the proximal arm of the male gonad show a positive reaction indicating the presence of this protein. By immunocytochemical localization in spermatozoa, MSP appears to be concentrated in the fibrous bodies in spermatocytes and ultimately in the pseudopodia, implying a possible motility function in the amoeboid sperm.[54]

Pulse-labeling experiments demonstrate that MSP is first synthesized at 39 to 42 hr after hatching at 20°, a time which corresponds to the early stages of spermatogenesis in the male nematode. This also corresponds to the earliest detection of poly A mRNA coding for MSP by in vitro translation.[55] This last result suggests that the regulation of MSP is controlled at the transcriptional level.

Experiments with actinomycin-D and α-amanitin indicated that transcription was required at approximately 35 to 40 hr after hatching. Furthermore, synthesis of MSP mRNA begins at 35 hr after hatching during the early L4 stage as demonstrated by Northern blot analysis using a genomic clone containing the MSP gene as a probe.[57] No MSP RNA transcript is detected before this time. Finally, the cell-specific nature of MSP gene expression was demonstrated by the *in situ* localization of MSP mRNA to a specific region of the gonad by *in situ* hybridization. For example, Figure 5 shows an autoradiograph of an isolated male gonad showing a concentration of grains above the mid-proximal arm. No other tissue showed any hybridization to the probe. Details of this procedure have been published elsewhere.[57] Such *in situ* hybridization techniques are currently being used to localize mRNA transcripts from a number of different genes.[58]

With the advent of molecular cloning techniques it has become routine to isolate stage- or sex-specific genes. For example, male-specific genes have been isolated from *C. elegans* in the following fashion. Poly A mRNA was isolated from male-enriched cultures of 80% adult male nematodes. This poly A mRNA was cloned using cDNA cloning techniques.[71] The resulting cDNA clone bank was screened by a differential hybridization technique. Labeled cDNA was made from poly A mRNA from adult males and used to probe a copy of the cDNA clone bank. A second cDNA probe was made from poly A mRNA from gravid (postspermatogenic) hermaphrodites and was used to probe a second copy of the cDNA bank. Those colonies showing hybridization only to the male cDNA probe and not to the hermaphrodite probe were selected as candidates for containing genes expressed only in the

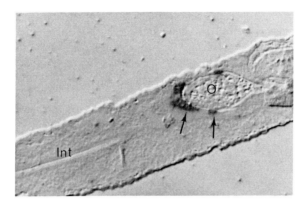

FIGURE 4. This is a photograph of a longitudinal section of a hermaphrodite taken with interference-contrast optics and labeled with antibody to MSP by the peroxidase/antiperoxidase method. The arrows indicate sperm in the spermatheca surrounding an oocyte (O). Also visible is the lumen of the intestine (Int).

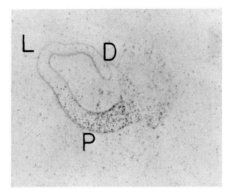

FIGURE 5. This is a light microscope picture of an isolated male gonad after *in situ* hybridization and autoradiography as described in the text. Grains represent regions of hybridization indicating presence of MSP mRNA. (D) distal tip; (L) loop region; (P) proximal arm.

male. Of 45 clones selected in this manner, 32 were shown to contain copies of the gene for the MSP. Whole genomic blots as well as analysis of genomic clone banks have shown that there are 15 to 30 copies of the MSP genes in the genome. The entire nucleotide sequence of the MSP gene has been obtained showing a probable promoter region (TATA box) 64 bp 5' to the initiation codon as well as a possible ribosome binding site and AATAAA polyadenylation site on the 3' trailing sequence. The MSP gene family is stable during development and shows no rearrangement or amplification in either the male or hermaphrodite.[72]

Utilizing the sperm isolation technique and the cloned genes, spermatogenesis-defective mutants isolated and described by Hirsh and Vanderslice[37] and Ward et al.[38,39,59] can now be analyzed for specific molecular and biochemical defects. These studies will provide information on the genetic control of spermatogenesis at the molecular level and may provide insight into the mechanisms of cell-specific gene expression.

One interesting aspect of spermatogenesis that changes with age is sperm fertility. If males of different ages are mated with spermatogenesis-defective hermaphrodites in such a manner that all of the fertilized eggs produced are fertilized by the male sperm, there is a significant reduction in the viability of the fertilized eggs produced. Also, the sperm produced by aging males exhibit a decline in their ability to fertilize oocytes (see Figure 5A).

## C. Hypodermis

Another well-documented case of developmentally regulated cell-specific gene expression is that of the collagen genes. The cuticle of *C. elegans* is a complex structure composed primarily of collagen synthesized by the hypodermal cells.[65-66] The collagen genes are developmentally regulated coincident with the periodic replacement of the cuticle during growth and maturation.[65] A large number of collagen species have been identified and collagens specific to different developmental stages have been characterized.[67] The *C. elegans* genome contains 50 or more collagen genes, 2 of which have been sequenced in their entirety.[68] Mutants with defective cuticles have been isolated and are currently being used for the genetic analysis of cuticle synthesis and assembly.[69-70]

## D. Muscle

Other investigators are using the nematode to analyze the genetic and biochemical aspects of myosin production.[19-24,60,61] Although myosin appears to be synthesized in small amounts by many cell types, it is found in a high concentration in the muscle cells and in this sense can be viewed as a specific cellular phenotype. In the nematode the muscle cells include the body-wall musculature and the bilobed pharynx. Through the exploitation of two muscle mutants, Epstein et al.[56,60,61] have been able to demonstrate the existence of two different structural genes for myosin heavy chains.

One mutant, CB675, contains a small deletion in one of the structural genes (the *unc-54* gene) for heavy-chain myosin and produces a 203,000-dalton heavy chain in addition to the normal 210,000-dalton chain. A second mutant, CB190, fails to produce the *unc-54* gene product. The investigators were able to obtain antibody-specific for the *unc-54* myosin by making an affinity column using myosin from the mutant which lacks the *unc-54* myosin and absorbing all the myosin antibodies except those specific to the *unc-54* myosin. Using this specific anti-*unc-54* myosin antibody Mackenzie et al.[56] were able to localize the two different myosins in the same body-wall muscle cells.

Whole worms fixed in 4% formaldehyde and embedded in polyethylene glycol 4000 were cut and used for immunocytochemical localization of the two myosins. Using the peroxidase/anti-peroxidase method and anti-myosin antibody (Figure 6), it was possible to detect myosin in all body-wall and pharyngeal muscle cells. Myosin was localized to the birefringent A-bands that contain thick filaments in the body-wall muscle. Secondly, using the purified anti-*unc-54* myosin, it was shown that *unc-54* myosin is present in all body-wall A-bands uniformly but is not present in the pharynx (Figure 6B). These two types of myosin may be coordinately synthesized.[20] The structural significance of two different myosins in the same muscle filament lattice is now being investigated at both the genetic and biochemical level. Most recently, the gene for the *unc-54* myosin heavy chain has been cloned.[21,22] This should facilitate the analysis of the regulation of this gene during development.

## E. Intestine

Another interesting cellular phenotype that is being used to study differential gene activity in the nematode is intestinal cell fluorescence. The intestinal cells of *C. elegans* contain autofluorescent compounds that can be observed microscopically and can be separated by thin layer chromatography. Babu,[62,63] using a high-pressure mercury lamp and a dissecting microscope, screened the F2 progeny from mutagenized worms for variance in fluorescence.

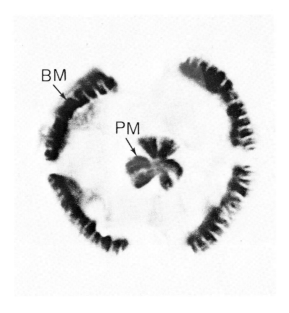

A

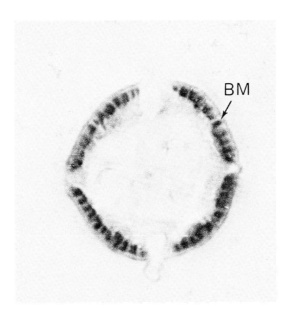

B

FIGURE 6. (A) This figure shows a cross-section of an adult hermaphrodite under bright field optics stained with antibody to myosin
using the peroxidase/antiperoxidase method. Both the body-wall musculature (BM) and the muscles of the pharynx (PM) show a positive
reaction. (B) This is a cross-section of an adult in the head region
stained with antibody specific for the *unc-54* myosin only, using the
peroxidase method. Only the body-wall musculature shows a positive
reaction.

At least 15 mutants defining 4 complementation groups, each with a specific fluorescence phenotype, have been isolated. At least two of these complementation groups have been shown to be involved in kynurenine catabolism, one affecting kynurenine hydroxylase (*flu-1*) and the other kynureninase (*flu-2*).

The intestinal cell fluorescence has also been used by Laufer et al.[64] as a cell-specific differentiation marker to determine the effect of blocking cell division in early embryos on the appearance of fluorescence. Using colchicine and cytochalasin B to inhibit cell division, they have shown that fluorescent pigment does appear at approximately the same time and place as in nonblocked embryos, indicating that some segregation of developmental potential occurs during the first cleavage stages.

We have only mentioned a few of the cell-specific phenotypes of *C. elegans* that are currently being investigated. These studies demonstrate the opportunity *C. elegans* offers for a multidisciplinary approach to the study of cellular differentiation. The advantages of being able to perform both genetic and biochemical analyses in a simple system whose entire anatomy is known at the cellular level will hopefully provide a better understanding of differential gene expression, and ultimately of development and aging.

# REFERENCES

1. **Brenner, S.,** The genetics of *Caenorhabditis elegans, Genetics,* 77, 71, 1974.
2. **Dougherty, E. C., Hansen, E. L., Nicholas, W. L., Mollett, J. A., and Yarwood, E. A.,** Axenic cultivation of *Caenorhabditis elegans* (Nematode: Rhabditidae) with supplemented and unsupplemented chemically defined media, *Ann. N.Y. Acad. Sci.,* 77, 176, 1959.
3. **Rothstein, M.,** Practical methods for the axenic culture of the free-living nematodes *Turbatrix aceti* and *Caenorhabditis briggsae, Comp. Biochem. Physiol.,* 49B, 669, 1974.
4. **Gandhi, S., Santelli, J., Mitchell, D. H., Stiles, J. W., and Sanadi, D. R.,** A simple method for maintaining large, aging populations of *Caenorhabditis elegans, Mech. Aging Dev.,* 12, 137, 1980.
5. **Sulston, J. E., Schierenberg, E., White, J. G., and Thompson, J. N.,** The embryonic cell lineage of the nematode *Caenorhabditis elegans, Dev. Biol.,* 100, 64, 1983.
6. **Kimble, J. and Hirsh, D.,** The postembryonic cell lineages of the hermaphrodite and male gonads in *Caenorhabditis elegans, Dev. Biol.,* 70, 396, 1979.
7. **Deppe, U., Schierenberg, E., Cole, T., Krieg, C., Schmitt, D., Yoder, B., and von Ehrenstein, G.,** Cell lineages of the embryo of the nematode *Caenorhabditis elegans, Proc. Natl. Acad. Sci., U.S.A.,* 75, 376, 1978.
8. **Von Ehrenstein, G. and Schierenberg, E.,** Cell lineages and development of *C. elegans* and other nematodes, in *Nematodes as Biological Models,* Vol. 1, Zuckerman, B., Ed., Academic Press, N.Y., 1980, 1.
9. **Cassada, R. C. and Russell, R. L.,** The dauer larva, a post-embryonic developmental variant of the nematode *Caenorhabditis elegans, Dev. Biol.,* 46, 326, 1975.
10. **Klass, M. and Hirsh, D.,** Non-aging developmental variant of *Caenorhabditis elegans, Nature (London),* 260, 523, 1976.
11. **Hirsh, D., Oppenheim, D., and Klass, M.,** Development of the reproductive system of *Caenorhabditis elegans, Dev. Biol.,* 49, 200, 1976.
12. **Josse, J. and Harrington, W. F.,** Role of pyrolidine residues in the structure and stabilization of collagen, *J. Mol. Biol.,* 9, 269, 1964.
13. **Bird, A. F. and Dutsch, K.,** The structure of the cuticle of *Ascaris lumbricoides var. Siris, Parasitology,* 47, 319, 1957.
14. **Ward, S., Thomas, N., White, J. G., and Brenner, S.,** Electron microscopical reconstruction of the anterior sensory anatomy of the nematode *Caenorhabditis elegans, J. Comp. Neurol.,* 160, 313, 1975.
15. **Ware, R. W., Clark, D., Crossland, K., and Russell, R. L.,** The nerve ring of the nematode *Caenorhabditis elegans:* sensory input and motor output, *J. Comp. Neurol.,* 162, 71, 1975.

16. **White, J. G., Southgate, E., Thomson, J. N., and Brenner, S.,** The structure of the ventral nerve cord of *Caenorhabditis elegans, Phil. Trans. R. Soc. London*, B275, 327, 1976.

17. **Lewis, J. A. and Hodgkin, J. A.,** Specific neuroanatomical changes in chemosensory mutants of the nematode *C. elegans, Neurology*, 172, 489, 1977.

18. **Albertson, D. G. and Thomson, J. N.,** The pharynx of *C. elegans, Phil. Trans. R. Soc. Lond.*, B275, 300, 1976.

19. **Epstein, H. F., Schachat, F. H., and Wolff, J. A.,** Molecular genetics of nematode myosin, in *Pathogenesis of Human Muscular Dystrophies,* L. P. Rowland, Ed., Excerpts Medica, Amsterdam, 1977, 460.

20. **Garcea, R. L., Schachat, F., and Epstein, H. F.,** Coordinate synthesis of 2 myosins in wild-type and mutant nematode muscle during larval development, *Cell,* 15, 421, 1978.

21. **MacLeod, D., Waterston, R. H., Fishpool, R. M., and Brenner, S.,** Identification of the structural gene for the myosin heavy chain in *C. elegans, J. Mol. Biol.,* 114, 133, 1977.

22. **MacLeod, A. R., Karn, J., and Brenner, S.,** Molecular analysis of the unc-54 myosin heavy-chain gene of *C. elegans, Nature (London),* 290, 386, 1981.

23. **Harris, H. E. and Epstein, H. F.,** Myosin and paramyosin from *C. elegans:* biochemical and structural properties of normal and mutant proteins, *Cell,* 10, 709, 1977.

24. **Harris, H. E., Tos, Man-Yin W., and Epstein, H. F.,** Actin and myosin-linked calcium regulation in the nematode *C. elegans.* Biochemical and structural properties of native filaments and purified proteins, *Biochemistry,* 16, 859, 1977.

25. **Klass, M. R.,** Aging in the nematode *C. elegans:* major biological and environmental factors influencing lifespan, *Mech. Aging Dev.,* 6, 413, 1977.

26. **Honda, H.,** Experimental and cytological studies on bisexual and hermaphrodite free-living nematodes, with special reference to problems of sex, *J. Morphol. Physiol.,* 40, 191, 1925.

27. **Klass, M., Wolf, N., and Hirsh, D.,** Development of the male reproductive system and sexual transformation in the nematode *Caenorhabditis elegans, Dev. Biol.,* 52, 1, 1976.

28. **Wolf, N., Hirsh, D., and McIntosh, J. R.,** Spermatogenesis in males of the free living nematode, *Caenorhabditis elegans, J. Ultrastruc. Res.,* 63, 155, 1978.

29. **Nigon, V.,** Les Modalities de la reproduction et le determinisme du sexe chez quelques nematodes libres, *Ann. Sci. Nat. Zool.,* 2, 1, 1949.

30. **Hodgkin, J. A., Horvitz, H. R., and Brenner, S.,** Nondisjunction mutants of the nematode *Caenorhabditis elegans, Genetics,* 91, 67, 1977.

31. **Hodgkin, J. A. and Brenner, S.,** Mutation causing transformation of sexual phenotype in the nematode *C. elegans, Genetics,* 86, 275, 1977.

32. **Herman, R. K.,** Crossover suppressors and balanced recessive lethals in *C. elegans, Genetics,* 88, 49, 1978.

33. **Herman, R. K., Albertson, D. G., and Brenner, S.,** Chromosome rearrangements in *Caenorhabditis elegans, Genetics,* 83, 91, 1976.

34. **Waterston, R. H. and Brenner, S.,** A suppressor mutation in the nematode acting on specific alleles of many genes, *Nature (London),* 275, 715, 1978.

35. **Waterston, R. H.,** A second informational suppressor, sup-7x, in *C. elegans, Genetics,* 97, 307, 1981.

36. **Herman, R. K. and Horvitz, H. R.,** Genetic analysis of *C. elegans,* in *Nematodes as Biological Models,* Vol. 1, Zuckerman, B., Ed., Academic Press, N.Y., 1980, 227.

37. **Hirsh, D. and Vanderslice, R.,** Temperature-sensitive developmental mutants of *Caenorhabditis elegans, Dev. Biol.,* 49, 220, 1976.

38. **Ward, S., Argon, Y., and Nelson, G. A.,** Sperm morphogenesis in wild-type and fertilization-defective mutants of *C. elegans, J. Cell Biol.,* 91, 26, 1981.

39. **Argon, Y. and Ward, S.,** *C. elegans* fertilization-defective mutants with abnormal sperm, *Genetics,* 96, 413, 1973.

40. **Ward, S.,** Chemotaxis by the nematode *C. elegans:* identification of attractants and analysis of the response by use of mutants, *Proc. Natl. Acad. Sci., U.S.A.,* 70, 817, 1973.

41. **Riddle, D. L.,** A genetic pathway for dauer larva formation in *C. elegans, Genetics,* 86, S51, 1977.

42. **Albertson, D. G., Sulston, J. E., and White, J. G.,** Cell cycling and DNA replication in a mutant blocked in cell division in the nematode *C. elegans, Dev. Biol.,* 63, 165, 1978.

43. **Vanderslice, R. and Hirsh, D.,** Temperature-sensitive zygote-defective mutants of *C. elegans, Dev. Biol.,* 49, 236, 1976.

44. **Edgar, R. S. and Wood, W. B.,** The nematode *C. elegans,* a new organism for intensive biological study, *Science,* 198, 1285, 1977.

45. **Wood, W. B.,** Summary of workshop on nematodes. *Molecular Approaches to Eucaryotic Genetic Systems,* Vol. 8, Wilcox, G., Abelson, J., and Fox, C. F., Eds., Academic Press., N.Y., 1977, 357.

46. **Klass, M., Wolf, N., and Hirsh, D.,** Further characterization of a temperature-sensitive transformation mutant in *C. elegans, Dev. Biol.,* 69, 329, 1979.

47. **Sharrock, W.,** Yolk proteins of *C. elegans, Dev. Biol.,* 96, 182, 1983.

48. **Kimble, J. and Sharrock, W.,** Tissue-specific protein synthesis in *C. elegans, Dev. Biol.,* 96, 189, 1983.

49. **Foor, W. E.,** Zygote formation in *Ascaris lumbricoides, J. Cell. Biol.,* 39, 119, 1968.

50. **Foor, W. E., Johnson, M. H., and Beaver, P, C.,** Morphological changes in the spermatozoa of *Dipetalonemus vitaea in utero, J. Parasitol.,* 57, 1163, 1971.

51. **Ward, S. and Carrel, J. S.,** Fertilization and sperm competition in the nematode *C. elegans, Dev. Biol.,* 73, 304, 1979.

52. **Roberts, T. M. and Ward, S.,** Membrane flow during nematode spermiogenesis, *J. Cell Biol.,* 92, 113, 1982.

53. **Nelson, G. A., Roberts, T. M., and Ward, S.,** *C. elegans* spermatozoan locomotion: amoeboid movement with almost no actin, *J. Cell Biol.,* 92, 121, 1982.

54. **Ward, S. and Klass, M.,** Localization of the major protein in *C. elegans* sperm and spermatocytes, *Dev. Biol.,* 92, 203, 1982.

55. **Klass, M. R. and Hirsh, D.,** Sperm isolation and biochemical analysis of the major sperm protein and *C. elegans, Dev. Biol.,* 84, 299, 1981.

56. **Mackenzie, J. M., Jr., Schachat, F., and Epstein, H. F.,** Immunocytochemical localization of 2 myosins within the same muscle cells in *C. elegans, Cell,* 15, 413, 1978.

57. **Klass, M., Dow, B., and Herndon, M.,** Cell-Specific transcriptional regulation of the major sperm protein in *C. elegans, Dev. Biol.,* 93, 152, 1982.

58. **Ewards, M. K., Williams, D. L., and Wood, W. B.,** Detection of specific RNA sequences in oocytes and early embryos of *C. elegans* by *in situ* hybridization, *Cell Biol.,* 83 (2, pt. 2), 216A, 1979.

59. **Ward, S. and Miwa, J.,** Characterization of temperature-sensitive fertilization-defective mutants of the nematode *C. elegans, Genetics,* 88, 285, 1978.

60. **Schachat, F. H., Harris, H. E., and Epstein, H. F.,** Two homogeneous myosins in body-wall muscle of *C. elegans, Cell,* 10, 721, 1977.

61. **Schachat, F. H., Harris, H. E., Garcea, R. L., La Pointe, J. W., and Epstein, H. F.,** Studies on two body-wall myosins in wild-type and mutant nematodes. Molecular approaches to eucaryotic genetic systems, *Symp. Mol. Cell. Biol.,* 8, 373, 1977.

62. **Babu, P.,** Biochemical genetics of *C. elegans, Mol. Gen. Genet.,* 135, 39, 1974.

63. **Siddiqui, S. S. and Babu, P.,** Genetic mosaics of *C. elegans:* a tissue-specific fluorescent mutant, *Science,* 210, 330, 1980.

64. **Laufer, J. S., Bazzicalupo, P., and Wood, W. B.,** Segregation of developmental potential in early embryos of *C. elegans, Cell,* 19, 569, 1980.

65. **Cox, G. N., Kusch, M., DeNevi, K., and Edgar, R. S.,** Temporal regulation of cuticle synthesis during development of *C. elegans, Dev. Biol.,* 84, 277, 1981.

66. **Cox, G. N., Kusch, M., and Edgar, R. S.,** Cuticle of *C. elegans:* its isolation and partial characterization, *J. Cell Biol.,* 90, 7, 1981.

67. **Cox, G. N., Staprons, S., and Edgar, R. S.,** The cuticle of *C. elegans.* II. Stage-specific changes in ultrastructure and protein composition during postembryonic development, *Dev. Biol.,* 86, 456, 1981.

68. **Kramer, J. M., Cox, G. N., and Hirsh, D.,** Comparisons of the complete sequences of two collagen genes from *C. elegans, Cell,* 30, 599, 1982.

69. **Higgins, B. J. and Hirsh, D.,** Roller mutants of the nematode *C. elegans, Mol. Gen. Genet.,* 150, 63, 1977.

70. **Cox, G. N., Laufer, J. S., Jusch, M., and Edgar, R. S.,** Genetic and phenotypic characterization of roller mutants of *C. elegans, Genetics,* 95, 317, 1980.

71. **Efstratiadis, A., Kafator, F. C., Maxam, A. M., and Maniatis, T.,** Enzymatic *in vitro* synthesis of globin genes, *Cell,* 7, 279, 1976.

72. **Klass, M., Kinsely, S., and Lopez, L.,** Isolation and characterization of a sperm-specific gene family in the nematode *Caenorhabditis elegans, Mol. Cell. Biol.,* in press.

73. **Blumenthal, T., Squire, M., Kirtland, S., Cane, J., Donegan, M., Spieth, J., and Sharrock, W.,** Cloning of a yolk protein gene family from *Caenorhabditis elegans, J. Mol. Biol.,* in press.

Chapter 4

# ANALYSIS OF THE BIOLOGICAL BASIS OF AGING IN THE NEMATODE, WITH SPECIAL EMPHASIS ON *CAENORHABDITIS ELEGANS*

**Thomas E. Johnson**

## TABLE OF CONTENTS

# I. INTRODUCTION

## A. Why Use the Nematode to Study Aging?

In the preparation of this review, I have made the basic assumption that the desire of the reader is to understand the biological basis of organismic aging. Given this premise, the organism of choice should be one that offers the most immediate hope of arriving at such an understanding. An ideal organism should have a short lifespan; be inexpensive to maintain; be experimentally malleable by a variety of techniques including molecular, morphological, genetic, and biological approaches; and be the object of study in a sufficient number of different laboratories to assure the accumulation of a critical mass of data. The nematode, *Caenorhabditis elegans,* admirably fulfills all of these basic requirements.[1]

Researchers in the field of aging are faced with a large number of different theories which purport to explain the molecular basis of organismic aging. There are two major reasons for this proliferation of theoretical views. First, aging is an extremely complex phenomenon involving changes in a number of different physiological systems; these physiological changes are often detected, but proof that any one of the changes is responsible for aging is lacking. Second, the focus of a great deal of the research in the field has not been so much on understanding the biological basis of the entire aging process as on understanding one or another of the consequences of this process, particularly in humans and in other mammals. The mammalian model systems may often be quite inappropriate for addressing the more basic, long-term questions about the nature of the primary aging process(es).

## B. The Power of Genetic Analysis

Genetic methods have contributed greatly to the rapid advances in our understanding of biological processes during the last 3 decades. In recent years, genetic approaches have been especially fruitful in the fields of developmental biology and behavioral biology. However, due primarily to a paucity of mutations which have specific effects only on the aging process, genetic techniques have rarely been used to investigate aging. This subject will be discussed at more length in Section III.H, where evidence which suggests that self-fertilizing organisms such as *C. elegans* are organisms of choice for the genetic analysis of aging will be reviewed.

The power of genetics lies in its ability to approach biological phenomena in a holistic and unprejudiced way. Selection for a mutant with a longer lifespan selects an alteration in a molecular process that limits length of life. The mutants themselves, therefore, suggest which processes are important in determining lifespan and which are irrelevant.

By far the most powerful approach in modern biology represents a combination of molecular and genetic techniques. Molecular analyses offer the possibility of penetrating insights into biological processes, but have little if any ability to determine the relative importance of these processes to the organism. However, when molecular techniques are coupled with the analysis of genetically altered strains, much more information concerning correlation and causality can be obtained. Proof of causality can even be established through the analysis of revertants, recombinants, etc. Thus, the judicious and concomitant use of the methods of genetics and molecular biology offers the possibility of analytical advances that could not be achieved by either approach alone.

Precisely such a combined genetic and molecular approach is being employed in *C. elegans* studies of development, behavior, and muscle physiology, as well as in studies of aging and other processes.

## C. Scope of This Review

This review is written with the intent of satisfying the needs of a variety of readers. Background on the biology, life cycle, and development of *C. elegans*, which can be found in Chapter 3 of this book and in other reviews,[1-6] receives little attention here. I will attempt to be both comprehensive and selective by covering a number of studies, but often not in great detail. Although primary emphasis is on *C. elegans*, other nematode species of importance in aging research, including *Turbatrix aceti* (the vinegar eel), *Caenorhabditis briggsae*, and *Panagrellus redivivus* are also mentioned. Unless the species is specifically identified, all references pertain to work on *C. elegans*.

## II. METHODS OF CULTURING AGING POPULATIONS

### A. The Problems

Two major problems must be overcome in maintaining synchronously aging populations of *C. elegans*. A synchronous group of young worms must somehow be obtained, and this group must then be maintained through life uncontaminated by the 200 or more progeny which each self-fertilizing, hermaphroditic nematode will produce. These problems have been overcome in a variety of ways. Aging cultures of *C. elegans* have been maintained over a temperature range of 10 to 26°C under several different culture conditions, both axenic and monoxenic. Axenic media, i.e., media containing no living organisms, include completely defined, undefined, and dead bacteria as food source, whereas monoxenic media contain living *E. coli*. Certain conditions are more suitable for preparation of mass cultures for biochemical analysis, while others are more suited to genetic studies.

The choice of an appropriate regimen for growth and maintenance of synchronously aging populations must therefore take into consideration stock maintenance and biological techniques, the method of synchronization, the type of media, the method of preventing progeny contamination, and the specific purpose of the study. These topics will be addressed separately. Three other reviews have approached the problem of maintaining synchronously aging populations from a somewhat different point of view.[7-9]

## B. Stock Maintenance and Biological Techniques

Modern studies of *C. elegans* were initiated by Brenner[10] in the late 1960s in an effort to investigate behavior through the use of genetic mutants and neurobiological procedures. His landmark paper describes a number of useful biological techniques and media which are used for growth, maintenance, and genetic analysis of *C. elegans*. Most laboratories currently use procedures similar to those originally developed by Brenner.

Of particular importance for the genetic study of aging is the fact that frozen stocks can be stored indefinitely in liquid nitrogen.[10] This frozen storage technique is supplemented by maintenance of master stocks prepared by allowing stocks of *C. elegans* to starve completely and form dauer larvae (see Section III.H.1) which are able to live for months[11] if the plates are protected from drying by wrapping in parafilm.[12]

Worms can be manipulated on solid media by the use of pointed sticks, pointed wire, or platinum spatulas; in liquid media, thin pipettes can be used. Unless otherwise noted, all studies reported herein used the standard *C. elegans* growth temperature of 20°C.

## C. Methods of Synchronization

Four different techniques have been employed to obtain synchronous young worms for starting cultures. The simplest is the use of eggs laid by young adult hermaphrodites over a short period of time. Other techniques useful for obtaining larger quantities of worms include hypochlorite treatment to isolate eggs uncontaminated by any living larvae or adults, an egg "hatch-off" to obtain larvae synchronized at the time of hatch, and "multiple screening" techniques which take advantage of the correlation between age and body diameter to separate young larvae from older larval forms and adults. Reference to Chapter 3 is recommended for readers interested in the complete understanding of the biological bases of these techniques.

### 1. Use of Eggs Laid by Young Adult Hermaphrodites

This technique is suitable for obtaining small quantities of synchronous eggs.[12-14] A synchronous population is obtained as follows: a plug of agar containing a few worms is transferred from a master stock to a fresh nutrient agar (NGM) plate (Figure 1). After 2 days, several young adult or fourth larval stage hermaphrodites are transferred to fresh NGM plates prespotted with *E. coli*. Fourth larval stage progeny are picked to fresh plates 5 days later and allowed to mature for 2 days before they are again transferred to a fresh plate where they lay eggs for a few hours before being removed and discarded. The eggs hatch and develop to adulthood in 3 days, and the adults are then transferred to a liquid medium that consists of S basal and *E. coli* at a concentration of $10^9$/m$\ell$.

This technique, which we use in our genetic analyses, provides a simple means of obtaining small quantities of synchronous eggs without harmful chemical treatment and with no possible effect of maternal age on lifespan or rate of senescence (see Section III.B.5). The procedure is not suitable for mass cultures of more than a few thousand worms.

### 2. Hypochlorite Treatment to Isolate Eggs for Mass Cultures

This technique[15] has been adopted by a number of researchers and also is used in our studies. A mixed population including a large number of fertile, well-fed adults is suspended by gentle rinsing of the agar plate with S basal.[10] The worms are pelleted and exposed to the hypochlorite solution until digestion begins releasing eggs from the bodies of the mothers. Digestion is continually monitored with a dissecting microscope, since the time of exposure to hypochlorite is critical and variable; overexposure can cause loss of viability. The eggs are quickly collected by pelleting and are thoroughly rinsed with S basal. Additional quantities of older eggs can be obtained by scraping the plates free of eggs laid before the mothers were isolated. Isolation of eggs from worms isolated by a gentle wash yields populations

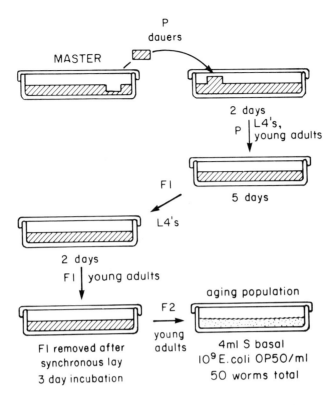

FIGURE 1.   Standard protocol for the establishment of small synchronous survival populations for use in genetic analyses of lifespan.[12,13]

of eggs that are usually less than 3 hr postfertilization. It should be noted that some genetic stocks are especially sensitive to this procedure and fail to yield viable eggs.

## 3. "Hatch-Off"

Larvae synchronized at the time of hatch are produced by this treatment.[16,17] Eggs are obtained as outlined in parts 1 or 2 above and are allowed to attach to a clean agar plate by air drying the surface after application. After about 8 hr S basal is gently washed over the surface to remove newly hatched larvae, leaving unhatched eggs still attached. Additional populations of newly hatched larvae can be obtained by successive washes of the plate.

## 4. Multiple Screening

Young larvae can be obtained by using a fine-mesh screen[8,18] similar to that used by Tilby and Moses.[19] A 13-μm mesh allows early larvae to pass, but retains older larval forms.

## 5. Final Points

The technique of choice is largely a matter of habit and personal taste. However, the method of achieving synchrony can significantly affect mean lifespan of the population. For example, we sometimes observe that hypochlorite treatment (technique 2) as opposed to synchronous egg lay (technique 1) produces worms that live longer. This may be due to the treatment itself or to slight variations in culture conditions such as density of bacterial growth during larval development which may result in caloric restriction. It should also be noted that maternal age effects on progeny lifespan have been reported (see Section III.B.5); thus, parental age should be controlled whenever possible.

## D. Types of Media

Several different types of media have been used for the maintenance of synchronously aging populations.[9,20] These include axenic media, containing either dead bacteria or chemically defined or undefined constituents, and monoxenic media that contain living *E. coli*. Monoxenic cultures have been maintained both in liquid media and in films of bacteria on the surface of agar plates.

### 1. Axenic Culture

Historically, the oldest techniques involve the use of axenic media.[9] The major advantages of this approach are the ability to control nutrients and to avoid metabolism by other living organisms. Disadvantages are that a large number of animals are killed by failure to lay eggs before the eggs hatch inside the body of the mother *(endotakia matricida, or "bags of worms")*. The fact that axenic media require sterile conditions adds difficulty to the performance of routine experiments. It is also clear that axenic media provide conditions quite unlike the growth conditions normally encountered by these nematodes in their natural environment, where they feed upon bacteria found in the soil.

Another type of axenic media is available that uses heat-killed or UV/heat-killed bacteria[21] as a food source. These approaches mimic more closely the native environment of the nematode while eliminating other living organisms from the system. They are appropriate for performing a number of physiological tests, such as measurement of oxygen uptake, etc. which would be quite difficult to do in the presence of living bacteria. However, preparation of dead bacteria is time-consuming and tricky.

### 2. Monoxenic Culture

These approaches to obtaining aging cultures are most closely analogous to those developed by Brenner.[10] The simplest method is to culture populations under conditions identical to those used by Brenner for maintaining stocks. This technique involves the use of NGM plates which have been prespotted with *OP50*, a uracil-requiring strain of *E. coli*. The Brenner approach has the advantage of ease of performance. Disadvantages are lack of rigid control of food concentration and ingestion, frequent contamination with other bacteria and with fungi, and the death of large numbers of nematodes by dessication on the sides of the petri plate.

We have chosen to use liquid culture media supplemented with living *E. coli* at $10^9$/mℓ,[13,22,23] a concentration which optimizes brood size[16] (see Section III.B.4 and Table 1). This technique permits easy handling through the use of drawn-out Pasteur pipettes and provides a good approximation of conditions encountered in the wild. Disadvantages are similar to those for axenic media except that only 10 to 15% of the hermaphrodites die as "bags of worms," perhaps because this culture medium provides more nutrients and thus avoids the egg retention and subsequent internal hatching associated with starvation.

## E. Methods of Preventing Progeny Contamination

*C. elegans* lays eggs from days 3 to 9 under standard liquid culture conditions.[12,16] Other related nematodes show similar but not identical patterns of reproduction during the first half of their normal lifespan. The adult synchronous populations being used for the aging studies must therefore be kept distinct from the progeny. Four major approaches have been used to keep progeny separate from adults: separation by a fine-mesh screen, use of FUdR for the sterilization of adult populations, manual separation of progeny and adults, and use of temperature-sensitive lethal mutations either to sterilize the adults or to prevent hatching of their progeny. The availability of four different methods of preventing progeny contamination provides an opportunity to compare results obtained by one technique with those obtained when one or more of the other approaches are used.

## Table 1
## EFFECT OF BACTERIAL
## CONCENTRATION ON LIFESPAN

| Bacterial concentration | Lifespan (days) Mean ± SE[a] | Number fertilized eggs |
|---|---|---|
| 0 | 4 ± 1 | 0 |
| $10^4$ | 5 ± 1 | 0 |
| $10^6$ | 5 ± 1 | 0 |
| $5 \times 10^7$ | 15.1 ± 1.5 | 14 |
| $10^8$ | 25.9 ± 2.4 | 63 |
| $5 \times 10^8$ | 19.4 ± 1.2 | 206 |
| $10^9$ | 16.0 ± 1.0 | 273 |
| $10^{10}$ | 15.0 ± 1.0 | 26 |

[a] Standard error.

### 1. Screening

Since young *C. elegans* have smaller diameters than do mature worms, one can use a screen of intermediate size to separate progeny and parents.[8,19] Metal or nylon meshes of appropriate diameter (approximately 20 μm) are arranged so that the aging population is suspended above the mesh in a layer of water thin enough to allow sufficient transpiration of oxygen. Additional media can be flushed through the apparatus until all young nematodes have been removed from the population above the mesh. This procedure is performed continuously or repeated periodically throughout the reproductive lifespan of the worm. Similar procedures using nylon mesh nets have been used to obtain populations of males separated from hermaphrodites by their smaller mature diameter.[24]

Advantages of this technique are that large populations can be maintained. Disadvantages include the problem of contamination by dead worm carcasses and ''bags of worms'' and the difficulty of completely removing all progeny from the population, which may explain the excessively long lifespans observed by Tilby and Moses.[19] However, this technique has been used to obtain large quantities of nematodes for use in studying DNA alterations in old worms.[25]

### 2. FUdR Sterilization of Adults

One of the easiest ways to avoid the problem of progeny contamination is to sterilize adults so that no progeny are produced. 5-Fluorodeoxyuridine (FUdR) has been used at concentrations varying from 25 μM to 30 mM to produce sterile adults and to prevent newly laid eggs from developing past the first larval stage.[8,9,14,17,18,26,27]

Precise techniques for the application of this method have been described.[17,26,27] The method has several advantages, including the fact that some conditions induced by FUdR treatment are reversible; for example, worms made sterile by FUdR recover fertility within 6 to 18 hr after removal from the drug,[27] a property that can be useful in mutant selections. Furthermore, at least some of the physiological modes of action of the drug are known. On the other hand, FUdR sterilization has been criticized since the treatment may severely affect the normal life history of the worm. Bolanowski et al.[14] observed that nematodes treated with 400 μM FUdR at 2.5 days have drastically altered lifespans and behavioral traits throughout life. The generality of this finding has been questioned by Gandhi et al.,[26] who suggest that the improper time of application of FUdR is most likely responsible for the results observed. They further suggest that judicious use of FUdR at a concentration of 25 μM causes little alteration of normal senescence. This argument is further bolstered by the finding of no difference in the levels of altered enzymes in old nematodes in the presence or absence of FUdR.[18]

Mitchell and Santelli[27] have shown that newly hatched worms mature in the presence of 30 μ*M* FUdR on a monoxenic medium. The worms attained normal adult length, formed apparently normal gonads, and produced normal numbers of fertilized eggs. The extensive cell division and differentiation involved in formation of the reproductive system apparently occurred normally.

Worms treated with 30 μ*M* FUdR as adults showed normal cessation of oocyte production, characteristic morphological and behavioral age-related changes, and a normal lifespan. Worms treated with 30 μ*M* FUdR as larvae died 25% sooner than controls, suggesting that some processes occurring in the larval stages were affected. The effects of FUdR treament on survival time and on behavioral traits will be discussed in more detail in Sections III.B—D. It appears that FUdR treatment on monoxenic media may be a useful regimen, especially as a complement to other artificial methods of preventing progeny contamination.

*3. Manual Separation*

This technique has been used both on agar surfaces[11,12,14,16] and in liquid culture.[13,22,23,25a] Worms are transferred daily from days 3 to 8 either by picking or pipetting to fresh media. Thrice weekly transfer is sufficient after the 8th day.[13] This method has the advantages of being technically easy and of providing a convenient means of performing behavioral and morphological assays, but has the obvious disadvantage of being quite labor intensive. This seems to be the method of choice for the genetic analysis of wild-type stocks.

*4. Use of Temperature-Sensitive Lethals*

Temperature-sensitive Ts mutants have been isolated in a number of laboratories. After induction by ethyl methane sulfonate, these mutants are isolated by their ability to grow at 16°C and their inability to grow at 25°C.[28,29] Mutants are available which are defective in a number of different processes, including sperm formation, gonad formation, embryonic development, sexual phenotype, and dauer development. Several Ts mutations have been used to block the production of viable progeny by aging adult populations by shifting the adults to the nonpermissive temperature at an appropriate time in their life. Two investigators have made use of a sperm-defective mutant strain (DH26) which fails to produce viable sperm at 25.5°C, but shows no measurable alteration in hermaphrodite lifespan.[12,16] We have also used embryonic mutations to block progeny development without measurable effects on lifespan[12] (see Section III.H.4).

The advantage of these approaches is that stocks can be sterilized in mass cultures with relative ease simply by varying temperature. Disadvantages include problems similar to those posed by drug treatments; i.e., heretofore unmeasured parameters may be affected by the mutation in question, and measurements of a mutant population consequently may not reflect true aging changes found in wild populations of animals. At least in some Ts mutants, however, the Ts function appears to be specific to certain dispensable tissues or to particular stages of life and therefore may not affect other adult functions. The use of several different Ts mutants is extremely attractive as one way of alleviating these problems, since different mutants are unlikely to respond in identical ways unless the processes under study do indeed have a significant effect upon length of life.

**F. Purpose of the Study**

One of the major variables in choosing culture conditions is the nature of the experiment. Studies of aging include analyses of morphological and behavioral changes in single worms or in small cultures, genetic studies of length of life in small numbers of worms, and biochemical studies on large populations.

The monoxenic approach appears to be most useful for genetic studies (liquid culture) and for morphological and behavioral analyses (solid or liquid culture). Monoxenic media most closely mimic the environment encountered by the worm in the wild and do not require the maintenance of absolute sterility. Small populations such as those used for genetic studies can be conveniently transferred by hand to minimize the problem of progeny contamination.

Large populations of worms for biochemical studies are most commonly prepared by axenic techniques using DNA synthesis inhibitors. Recently, however, mass cultures for isolation of nucleic acids from aged worms[25] and for analysis of the spectrum of proteins synthesized in old worms[30] have been accomplished through monoxenic culture of DH26, a ts spermatogenesis mutant. *T. aceti* has been cultured for the analysis of enzyme-specific activities in old worms both by the use of FUdR and by multiple screening.[18]

Ideally, a uniform technique which gives identical survival curves and other indices of senescence for single worm cultures, small population cultures, and mass cultures should be developed. It is likely that a variation of the monoxenic approach will be found to be most suitable for this purpose.[12]

## G. Comparison of Alternative Approaches for Maintenance of Aging Populations: The Debate

Many different techniques for maintaining synchronously aging populations of nematodes have been reviewed in this section. When the results of all of these studies are summarized, it becomes apparent that the survival curves obtained by the different methods stand in an orderly relationship to each other which can be intepreted as mainly reflecting variation in one major parameter, nutrition.

Lifespan determinations carried out at 20°C (excluding the inordinately short lifespan observed by Croll et al.[20]) vary from a mean of about 10 to 15 days on solid monoxenic media,[11,12,14,16,17,27,31] to a mean of about 18 days in liquid monoxenic[13] or axenic media using killed *E. coli*,[21] to a mean of 24 to 25 days in defined or undefined axenic media,[17,20,26] to a mean of 58 days for axenic growth coupled with screening for the avoidance of progeny contamination.[19] I suggest that the primary variable in this 5-fold variation in length of life is the amount of food ingested by the worm. Nematodes grown upon a solid agar substrate preseeded with *E. coli* live in an environment that is almost solid bacteria at concentrations near $10^{12}/m\ell$. At the lesser concentrations of liquid monoxenic culture ($10^9$ bacteria per milliliter), one would expect lengthened lifespans as demonstrated by Klass[16] (see Section III.B.4 and Table 1). Finally, axenic media would be expected to be the least nutritious due to the difficulty of artificially fulfilling the dietary requirements of a nematode. Starvation has been shown to increase the frequency of "bags of worms," which also become plentiful in axenic media.

It is apparent that one of the problems in the use of the nematode, especially *C. elegans*, as an aging model is that two schools of thought have come into conflict as to proper culture conditions. Those *C. elegans* researchers who have not been influenced by the Brenner school favor the semidefined axenic media. Those who come to the study of aging from the Brenner school of thought usually use monoxenic cultures or the closely allied technique of using heat-killed *E. coli* as a food source. The primary motivation of this latter group of researchers is to mimic normal environmental conditions as closely as possible. As already pointed out, different research questions require the use of different protocols. At least for some purposes, however, the variety of methods can serve as a strength by permitting cross-validation of results.

## III. AGING STUDIES

### A. Introduction

The earliest studies of aging in the nematode were performed prior to the introduction of *C. elegans* by Brenner. These studies used three other species of nematode: *Caenorhabditis briggsae, Panagrellus redivivus,* and *Turbatrix aceti* (the vinegar eel). Several of the studies covered in this section deal with these species exclusively.

Studies of aging are divided into seven categories in this review. Section B deals with analyses of lifespan and the effects of several biological variables on length of life. Sections C, D, and E focus primarily upon important background work which is descriptive in nature; the goal of these studies was to describe the major changes that occur at the morphological, behavioral, and molecular levels. More recent work has begun to seek answers to questions about the causes of aging and to test some models of the aging process; this research will be reviewed in Sections F, G, and H, which describe effects of food restriction and drug treatments, and discuss genetic studies of aging.

## B. Analyses of Lifespan
### 1. Criteria of Death and Data Analysis

In the determination of lifespan, some criteria must be established for determining whether any individual worm is living or dead. Bolanowski et al.[14] used four criteria: (1) lack of any spontaneous movement, (2) lack of pharyngeal pumping and defecation, (3) lack of any response to touch by a needle, and (4) degradation of normal gross morphology. Johnson and Wood[13] added the additional criterion of loss of body turgor pressure assayed by cutting those animals which have failed to pass the earlier criteria; living animals extrude body contents when cut.

Once the time of death is determined, the lifespan data must be analyzed. Most investigators either assume that time of death is normally distributed and then use statistics suitable for the analysis of normally distributed data or make no statistical tests at all. In contrast, Klass[16] used the nonparametric sign test in comparing populations of worms which had undergone different experimental procedures. Johnson and Wood[13] used two nonparametric statistics designed especially for survival analysis: the Gehan[32] (available in the survival subroutine of the Statistical Package for the Social Sciences),[33] and the log-rank test.[34] These statistics and others have been well described by Miller.[35] The use of statistical procedures derived especially for the analysis of survival data is probably preferable to assuming a normal distribution.

An additional advantage of the Gehan and log-rank survival analyses is that data on animals dying from causes other than natural ones can be discounted by statistically removing them from the data set at the appropriate time. Thus, Johnson and Wood[13] have three categories for recording the status of individual worms: alive, dead, and lost. "Lost" consists of individuals which were killed erroneously, which became "bags of worms," and which were really lost in transfer. We have further facilitated statistical analysis by the construction of computer software for direct storage of the data on a daily basis and by construction of software routines for the transformation of the data to a format compatible with the Gehan test and with our own software developed for the log-rank test. These programs are available from the authors.[36]

### 2. Survival Curves and Gompertz Plots of Males and Hermaphrodites

When data on the number of survivors at any given time after birth are plotted against age of the cohort of animals, the distribution is called a survival plot.[37] This plot assumes a characteristic form in the usual aging population. Survival curves for both male and hermaphroditic *C. elegans* wild-type stocks show this type of survival pattern (Figure 2A). The curves are sigmoidal, with an increasing probability of death as a function of age. In most, if not all animal species, the death rate increases exponentially with age in a relationship which has been called the Gompertz function.[37] Both male and hermaphroditic *C. elegans* have age-specific death rates that increase in this manner (Figure 2B). The slope of the regression of the natural logarithm of the death rate on age is a measure of the rate of

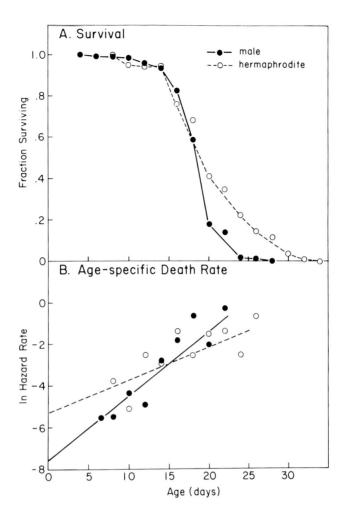

FIGURE 2. (A) Survival curves for male and hermaphroditic *C. elegans*.[12] (B) Age-specific death rate (hazard rate) was calculated by dividing the number of animals that died during a 2-day interval by the total size of the remaining survival population. The natural logarithm of this number was plotted throughout life, and linear regressions were determined. Slopes for the two curves are 0.307 and 0.153 for males and hermaphrodites, respectively; intercept points are −7.48 and −5.23, respectively.[12]

increase in mortality, and the Y-intercept of the regression line is an indication of the death rate at time zero. For the data on *C. elegans* (N2) males shown in Figure 2, these values are 0.307 and −7.48, respectively; the corresponding values for hermaphrodites are 0.153 and −5.23. Thus, hermaphrodites showed a higher probability of death early in life even though data on deaths not related to aging, such as those due to "bags of worms," have been removed from these analyses (see Section III.B.1). The increase in mortality with age however, occurs more slowly in hermaphrodites than in males; between 10 and 15 days of life, the male mortality rate becomes greater, and they eventually exhibit a shorter mean lifespan. Other nematode species show similar differences in mean lifespan between males and females, with the male lifespan typically being shorter.[9] Needless to say, other species of nematodes have very different mean lifespans;[9] for instance, Zuckerman et al.[38] showed that *C. briggsae* hermaphrodites at 22°C survive an average of 31 days.

Klass[25b] has found a mean lifespan of 14.4 days for *C. elegans* males and 18.7 days for hermaphrodites. In our own studies[12] we have determined the mean lifespans in populations of males and of hermaphrodites a number of different times over the period of a year. The average mean lifespan for males in these 12 studies was 14.9 days, whereas that for hermaphrodites was 18.5 days. The substantial variation in lifespan that was detected among experiments is examined in more detail in the next section.

*3. Consistency of Survival Measurements*

Average male lifespan in our 12 experiments[12] varied between 13.0 and 17.9 days, and the average lifespan of hermaphrodites varied between 16.1 days and 21.5 days. It should be noted, however, that variations in mean lifespan among experiments on other organisms are often much more extreme that those that we observed. If these variations in lifespan are primarily due to uncontrolled environmental effects at the different times the experiments were conducted (ambient temperature, etc.), we might expect the mean lifespans of the males and the hermaphrodites to covary. It is clear that male and hermaphrodite lifespans are related (Figure 3), showing better than 80% correlation and a highly significant regression of hermaphrodite lifespan on male lifespan. It is possible, therefore, that the differences in mean lifespan observed in different laboratories may be due to environmental effects beyond the control of the experimenter. The presence of such uncontrolled effects necessitates the inclusion of all appropriate control stocks in each lifespan determination.

In view of observed variations among different replications of the same measures, it is imperative to be sure that all genotypically identical populations within an experiment have identical survival curves. This problem has been addressed by Bolanowski et al.,[14] who found that several different populations which were used for different behavioral measurements nonetheless had similar lifespans. Reproducibility within an experiment was also analyzed by Johnson and Wood[13] in a study in which it was demonstrated that lifespan determinations were statistically reliable within an experiment. Several genetically identical control groups are routinely maintained in all of our lifespan determinations, and statistical analyses of consistency are performed before any comparisons are made between different populations.

*4. Effects of Temperature and Food*

Aging in poikilothermic organisms has been shown to be negatively related to temperature. *C. elegans* raised at 14°C live roughly twice as long as those raised at 25.5°C.[16] A linear relationship with temperature ($Q_{10}$ of about 2) is typical of biochemical reactions. Klass[16] has shown that there is a linear relationship between temperature and lifespan between 14 and 25.5°C (Table 2). Severe departures from linearity occur outside that range. The time of development is also proportionately lengthened at lower temperatures[16,39] (Table 1). The fact that a number of life-history traits are severely affected by temperatures below 14 or above 25.5°C indicates that *C. elegans* is maladapted for these temperature extremes.

Zuckerman et al.[38] have observed a significant difference in lifespan of *C. briggsae* at 17°C (30.0 days) and 27°C (26.2 days). Larval production was faster and began earlier in life at 27°C. Rates of movement showed linear decreases with age that occurred more rapidly at the higher temperature. Sensitivity to osmotic stress and formaldehyde was greater in older worms.

The concentration of bacteria in monoxenic cultures has been shown to have a large effect both on lifespan and on other life-history traits such as fecundity.[16] Maximum mean lifespan is achieved at $10^8$ bacteria per milliliter (Table 1), a concentration which decreases the number of progeny produced per hermaphrodite by over 65%. This large decrease in the number of progeny per worm appears to be characteristic of semistarvation conditions, and suggests that at this bacterial concentration the lives of the worms are being prolonged by effects of dietary restriction similar to those seen in experiments on higher organisms. A concentration of $10^9$ bacteria per milliliter produces maximum brood size and has conse-

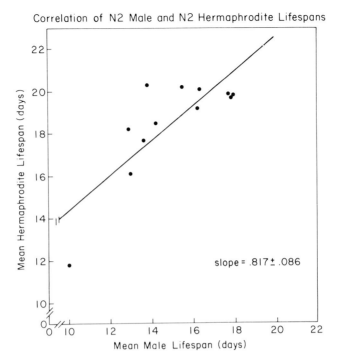

FIGURE 3.   Regression of mean hermaphrodite lifespan on mean male lifespan for 12 different determinations. Slope is 0.817 ± 0.086 (95% level of confidence).[12]

## Table 2
## EFFECTS OF TEMPERATURE ON LIFESPAN AND NUMBER OF FERTILIZED EGGS

| Temperature (°C) | Lifespan (days) Mean ± SE[a] | Number of fertilized eggs | Length of development (days) | Length of reproduction (days) |
|---|---|---|---|---|
| 6 | 17.8 ± 1.1 | 0 | — | — |
| 10 | 34.7 ± 3.0 | 84 | 10 | 14 |
| 14 | 20.8 ± 0.9 | 206 | 5 | 7.5 |
| 16 | 23.0 ± 1.6 | 250 | 4 | 7 |
| 20 | 14.5 ± 1.0 | 273 | 3 | 6 |
| 24 | 9.9 ± 0.4 | 269 | 2 | 4 |
| 25.5 | 8.9 ± 0.6 | 103 | 2 | 4 |

[a]   Standard error

quently been chosen as a standard condition by a number of laboratories including our own. When Klass shifted cultures from restricted ($10^8$ bacteria per milliliter) to unrestricted growth conditions and vice versa at various ages and examined lifespan, he observed a response in stocks shifted at any time. However, those shifted while still larvae showed the maximum prolongation of life.[16]

Croll et al.,[20] as well as Mitchell and Santelli,[27] found that monoxenic cultures had shorter lifespans and developed slightly faster than did cultures grown axenically. It appears that these findings are attributable to the dietary restriction that results from the use of axenic media. Although the results were originally interpreted as suggesting that monoxenic culture causes premature death by infection of aging worms,[20] this interpretation seems to have been ruled out by experiments in which worms which had aged in an axenic medium were transferred to monoxenic culture without a rapid increase in the rate of death after transfer.[17]

### 5. Parental Age Effects

Since Lansing's observations on rotifers, a number of studies of various organisms have investigated the possibility of a negative correlation between parental age and fecundity or lifespan of progeny. Beguet[40,41] showed that progeny of old worms had lower fecundity rates than did siblings produced by their parents earlier in life. These differences were not stable and no significant difference in number of progeny was seen after four successive generations of selecting for populations laid by old worms as compared to those laid by young worms. Klass[16] examined the lifespans of progeny produced by young nematodes (3 to 5 days old) and by older nematodes (6 to 9 days old) and made pairwise comparisons. Although the difference between mean lifespans of the progeny of young versus old worms was statistically significant, the average lifespans differed only slightly (14.23 and 13.86 days, respectively).

To the extent that the lengthened lifespan of longer-lived worms is a result of genetic factors, one would predict that longer-lived worms would produce progeny which themselves live longer. Klass[16] showed that the progeny of worms that lived only 7 to 8 days had a mean lifespan of 15.12 days, while the progeny of worms that lived 23 to 25 days had a mean lifespan of 17.83 days. Pairwise comparisons between progeny populations derived from parents with a variety of lifespans revealed that the correlation between parental and progeny lifespans was significant at the 0.01 level. This result is somewhat surprising because the autogamous hermaphroditic breeding structure of *C. elegans* yields genetic stocks that are 50% inbred at each generation, which should maintain almost complete genetic homozygosity. At face value this suggests that the observed differences in lifespan are not chromosomal, but are still "heritable."

## C. Morphological Studies

The first studies of aging in nematodes were primarily morphological and were performed mostly on *T. aceti* and *C. briggsae*. Early electron microscope (EM) studies of *C. briggsae* revealed a number of readily discernible changes as a function of age. The most striking and reproducible change is the accumulation of electron-dense material in the intestinal cells of aging *C. briggsae*[9,42] and *C. elegans*.[9] This also occurs to a lesser extent in *T. aceti*.[43] These inclusions appear early in life, develop throughout the normal lifespan, and eventually occupy most of the intracellular space of the intestine. The electron-dense inclusions show an association with acid phosphatase activity and sometimes show evidence of membranous whorls reminiscent of those seen in ceroid bodies.[42] The location of these inclusions and their morphology led Zuckerman and Himmelhoch[9] to the conclusion that they are lipofuscin in nature. As these inclusions come to fill most of the volume of the intestinal cells, they are associated with destruction of the cellular mitochondria, dissolution of the microvilli, and the disappearance of the endoplasmic reticulum.[42] Lipid droplets also accumulate in the intestine, but to a lesser extent.

Muscle cells of 35-day-old *C. briggsae* (mean half-life of 34 days) also show severe amounts of degeneration.[42] Old animals have fewer myofibrils, which tend to be more disorganized than those of young adults. The syncytial hypodermal cells show electron-dense inclusions and bulges,[9] as well as degenerating mitochondria and lysosome-like bodies[42] in the interchordal regions.

The seven-layered cuticle which surrounds the nematode body has been observed to accumulate an electron-dense material.[9,44] The fiber-filled layer also becomes difficult to distinguish. Scanning EM studies suggest that the cuticle wrinkles in old age, that the mouth closes, and that the deirids are difficult to see due to closing of the cuticle over them.[45]

Himmelhoch et al.[46] have observed less binding of cationized ferritin in old animals, which they interpret to mean that there is a 25% decrease in the net positive surface charge on the cuticle of these older worms.

EM studies of *T. aceti* have found age-related changes similar to those observed in *C. briggsae*.[43] The interchordal hypodermal regions thicken with advancing age, electron-dense inclusions appear within the pseudocoelum, and mitochondria appear to lose their inner cristae.

One problem of EM analysis is possible artifacts caused by differential fixation due to permeability changes in old worms. Such observations as wrinkling of the cuticle and the closing of cuticular pores are reminiscent of fixation artifacts and therefore must be replicated before they are accepted as valid correlates of aging.

For the most part these sorts of descriptive morphological studies have not been carried out in *C. elegans*, although available data generally suggest that these morphological changes are likely to occur in *C. elegans* as well.[9] *C. elegans* shows the accumulation of electron-dense inclusions in the intestinal cells and inclusions in the pseudocoelum.[9] Croll et al.[20] also observed 20% increases in length and volume of 14-day-old adults as compared with 5- or 7-day-old adults; however, this difference in volume is small in comparison with the size changes during larval development.

In order to be really useful in the study of the biological basis of aging, these types of morphological changes must be described quantitatively. In particular, it is important to determine the volume and relative proportion of the total intestinal cytoplasm that is occupied by lipofuscin inclusions. It is the opinion of this writer that the primary usefulness of morphological descriptions is as biological markers of aging and that the most important information concerning the nature of the aging process(es) will be obtained from quantitative studies of animals at different ages.

## D. Behavioral Changes
### 1. Movement and Chemotaxis

The nematode has a complex yet easily defined and quantifiable series of behaviors.[47,48] These include backwardly directed, whole-body sinusoidal waves used primarily for movements in the forward direction and forwardly directed waves used for backward movement. Changes in direction of swim are accomplished by an "omega" turn in which the worm bends back almost to its tail and then straightens out to begin moving in a new direction. The head of the worm has more complex and less defined movements which appear to constitute "searching" behavior for food. Complex movements of the head are powered by an anterior series of nerve and muscle cells, while the whole-body movements depend primarily upon alternate contraction and relaxation of muscles on either side of the body. Other observable behaviors have not been explained even at this descriptive level.

Several groups have examined age-related changes in whole body waves. Croll et al.[20] reported reduction in wave frequency with age for worms from 8 to 20 days old. Bolanowski et al.[14] found linear decreases in the frequency of whole body waves over an age range from day 4 to day 18, with an average decrease of 4.66 waves per minute per day (Figure 4).

FUdR sterilization has been used in a series of experiments to determine whether this treatment affects behavioral changes normally seen during aging.[14,17,26,27] Treatment with FUdR at 400 $\mu M$ resulted in threefold decreases in the frequency of whole body waves at all points in the lifespan[14] (Figure 4).

Hosono[49] reported age-related changes in the number of animals which chemotax to a patch of *E. coli*. He saw a 4-fold decrease over the first 4 days of life in the rate of accumulation of animals. Johnson[12] also observed a decline in the rate of taxis using another assay system.[50] One problem with both of these studies is that the ability to chemotax is not separated from the ability to move, which is known to decrease over this time period.

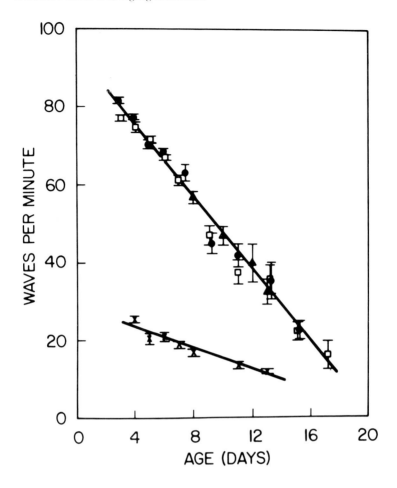

FIGURE 4. Population mean movement wave frequency ($\pm$ standard error) as a function of age. (▲), population A (only day 8, 10, 12, and 13 data shown for clarity); (●), population B; (□), population C; (×), FUdR-treated population. Lines represent linear regressions of the FUdR-treated and untreated population means. (From Bolanowski, M. A., Russell, R. L., and Jacobson, L. A., *Mech. Ageing Dev.*, 15, 279, 1981. With permission.)

Klass[25] observed a decrease in the ability of males to mate and produce viable offspring as a function of age. Using the temperature-sensitive, sperm-defective mutant strain (DH26), which can be rescued by mating with wild-type males carrying good sperm, he found that morphologically fertilized eggs showed successively higher rates of nondevelopment as the males aged (Figure 5A); hermaphrodite age was held constant. In another study of male mating behavior, Johnson[12] found that males lost the ability to mate at day 11 (Figure 5B).

## 2. Feeding

Rate of food ingestion can be readily monitored under a dissecting microscope by counting the frequency of pharyngeal contractions or, alternatively, the frequency of defecations. Croll et al.[20] observed that the rate of pharyngeal contractions in monoxenically grown worms was relatively constant until day 10, when a rapid decrease in rate began to occur which ended with death of the nematode population. Axenically grown worms showed almost no decrease in pump rates before day 18.[20] However, in two unpublished studies[12,27] the observations of Croll et al. on monoxenically grown cultures could not be repeated. Instead, maximal pump rates were seen in 3-day-old adults. These pump rates declined markedly

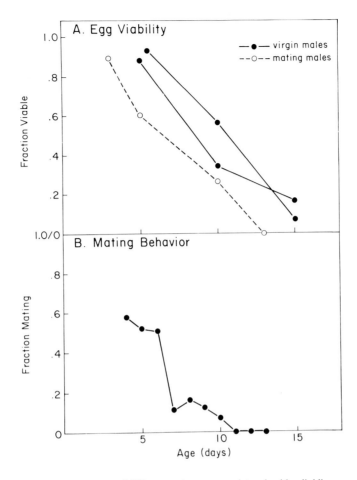

FIGURE 5. (A) Egg viability at various ages as determined by dividing the number of hatched eggs by the total number of eggs laid.[25] (B) Mating was assessed by calculating the fraction of males that exhibited mating behavior in a 10-min period.

from day 3 to day 6 and showed gradual declines thereafter. Gandhi et al.[26] observed little change in rate of pharyngeal pumping up to 14 days in axenic cultures and a rapid decrease thereafter that culminated in no pumping by day 30. In the same study, similar rates were observed in pump rates in populations of worms treated with low concentrations of FUdR (25 μM).

In a study by Bolanowski et al.,[14] defecation rates were found to decrease with age according to a function that showed three components: a rapidly decreasing early component from days 3 through 6 (−0.233 defecations per minute per day), a stable period, and a linear fall from days 9 to 14 (−0.089 defecations per minute per day). The authors pointed out that the late-onset linear decrease in rate was correlated with the cessation of egg-laying.

### 3. Fecund Period

Klass[16] found a linear decrease in length of the fecund period with temperature over the range of 10 to 25.5°C (Table 2). This decline with temperature is quite close to twofold per 10°C. Culture conditions have been observed to affect brood size, with maximum number of progeny obtained at $10^9$ bacteria per mℓ (Table 1). Gandhi et al.[26] have reported slight changes in the number of eggs produced in low concentrations of FUdR, but no change in the time of onset of reproduction. Croll[51] demonstrated a clear influence of age on vaginal/

vulval response to exogenously applied 100 μg/mℓ serotonin; both the percent of vulval contractions and the rate of contraction decreased dramatically between days 6 and 14. In *T. aceti*, Kisiel and Zuckerman[6,52] reported that a shorter fecund period was observed when older females were initially mated than when young females were mated.

### 4. Use of Behavioral Changes in Predicting Lifespan

Hosono et al.[53] observed that worms exhibited three distinct types of movement as they senesce. Younger worms displayed active, vigorous movements (type I), which became more irregular later in life (type II); finally, movements became progressively more irregular until the worms responded only when prodded (type III). Mean lifespans of these three types were 7 to 8, 3 to 4, and 1 to 2 days, respectively. The three types showed differential survival in response to 0.1 or 1.0% Nile Blue; all worms of type I, but none of type III, survived treatment with 1.0% Nile Blue. Bolanowski et al.[14] found a high correlation between movement rate and age in individual worms ($r = 0.87$). They then asked if the observed individual variability in decrease of movement could be used as a predictor of lifespan for individual worms, but found no correlation between rate of decline in movement and observed lifespan (Figure 6A). Furthermore, neither projected time of zero movement nor projected movement at day 0 was correlated with the expected lifespan of the individual worm (Figure 6B,C).

Klass[25a] has obtained a series of long-lived mutants induced by EMS mutagenesis (see Section III.H.5). All of these mutant strains have significantly altered rates of food ingestion, as determined by the rate of pharyngeal contraction,[25a] and they show alterations in rate of movement and other behaviors as well.[12] When Klass[16] investigated the relationship between lifespan and number of eggs laid by individual worms, he found no significant correlation.

### E. Molecular Studies

Most of the studies using nematodes as molecular models in aging research have been descriptive. These include analyses of age-related changes in activity of a number of different enzymes, in specific gravity, in rate of water efflux, and in the accumulation of cGMP and lipofuscin.

Erlanger and Gershon,[54] using FUdR-synchronized *T. aceti*, showed that acetylcholinesterase and α-amylase activities are maximal in a day 5 culture, that malic dehydrogenase and acid phosphatase activities peak at about day 15, and that acid ribonuclease activity is highest at day 5 and shows a second peak at day 35. There also appear to be changes in the isozymes of acid phosphatase at different ages as analyzed on neutral acrylamide gels.

In an excellent study, Bolanowski et al.[55] followed the activities of four different lysosomal enzymes: acid phosphatase, β-N-acetyl-D-glucosaminidase, β-D-glucosidase, and α-D-mannosidase. Using conditions under which the mean lifespan of the DH26 strain was 8.9 days, they assayed daily for enzymatic changes. From day 3 to day 10, they found a 2.5-fold increase in acid phosphatase levels, an 8-fold increase for β-N-acetyl-D-glucosaminidase, a 9-fold increase for β-D-glucosidase, and a 4-fold increase for α-D-mannosidase. Three forms of acid phosphatase and two forms of β-D-glucosidase were separated by ion exchange chromatography, and the observed increases in activity were found to be due primarily to an 18-fold increase in acid phosphatase I and a 100-fold increase in β-D-glucosidase I. Little change in activity of α-D-glucosidase, choline acetyltransferase, or acetylcholinesterase was detected, indicating that not all acid hydrolases show activity changes and that the decrease in movement seen in aged worms[14] is not attributable to the effect of lower levels of the enzymes involved in acetylcholine metabolism.

Willett et al.[56] found much higher cGMP concentrations in older cultures of *P. redivivus*, with concentrations increasing from about 70 p*M*/mℓ of medium in 4- to 10-day-old worms

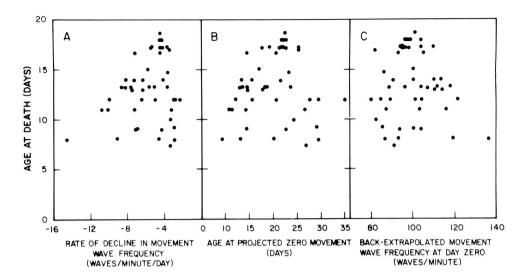

FIGURE 6.   Relationship between movement wave frequency and age at death of individual animals. (A) Rate of decline in movement wave frequency (waves/minute/day); (B) age at projected zero movement (days); (C) back-extrapolated movement wave frequency at day zero (waves/minute). Data are for individual animals of movement populations A, B, and C. (From Bolanowski, M. A., Russell, R. L., and Jacobson, L. A., *Mech. Ageing Dev.*, 15, 279, 1981. With permission.)

to 232 p$M$/m$\ell$ in 22-day-old worms. Klass[16] showed that a fluorescent compound similar in its excitation properties to lipofuscin accumulates in aging *C. elegans* to levels 8- to 20-fold higher in 22-day-old worms than in 2-day-old worms. Link et al.[57] have separated and tentatively identified three classes of substances responsible for this fluorescence: indoles, flavins, and lipofuscins.

Changes in specific gravity with age have been observed by Zuckerman et al.[58] Old worms have a specific gravity near 1.13, while that in young worms is near 1.05. Searcy et al.[59] reported a difference in water exchange rate across the cuticle between older (21-day-old) and younger (7-day-old) *C. briggsae*. The generality of this finding has been questioned by studies[14] in *C. elegans* which suggest that there is no change in the rate of $H_2O$ efflux over comparable periods of life. However, water exchange does appear to consist of two components, one fast ($t_{1/2}$ of 2.1 min) and one slower ($t_{1/2}$ of about 15 min).

McCaffrey and Johnson[30] used two-dimensional gel electrophoresis to examine the possibility of change in the spectrum of proteins synthesized in old (19-day) *C. elegans* as compared to those synthesized in young nematodes. They found no changes in the 200 or so most prominent bands. These were the only bands that could be detected in 19-day-old worms due to the decreased rate of incorporation of $^{35}S$ methionine in older cultures; 11-day-old worms also showed no changes in the approximately 900 spots resolvable at that age. The gels were also examined to see if there was any spreading of the bands in the isoelectric focusing dimension or if new satellite bands of major spots appeared later in life. Either finding would be an indication of the protein mistranslation that has been predicted by several error catastrophe models of aging. No evidence of mistranslation was detected. Hosono[49] reported myosin alterations in 8-day-old *C. elegans*, but presented no data.

Klass[25] investigated the possibility of single-strand nicks in the DNA of 15-day-old worms by using a Pol I-mediated assay to look for incorporation of $^3$H-TTP into isolated DNA. He found a 4-fold increase in this activity in 15-day-old as compared to 5-day-old worms.

A series of beautiful studies on the nature of the changes in specific activity observed in many different enzymes with age has been carried out by Sharma and Rothstein.[8,18,60-64] Using crude extracts of *T. aceti*, these investigators found that isocitrate lyase, phosphoglycerate kinase, and enolase showed age-related decreases in specific activity in older worms whose synchrony was maintained by successive screening or by FUdR treatment.[18] In particular, they have examined the decrease in specific activity of enolase in aging cultures of *T. aceti.* Old enolase is apparently composed of two different antigenic forms of the enzyme.[61] One of these is antigenically the same as "young" enzyme, while the other component appears to be an inactive form, which explains the decrease in specific activity. Inactive enzyme molecules of either young or old purified enolase are increased by chromatography on DE-52. Polyacylamide gel electrophoresis revealed differential mobility of the active and inactive forms. Side group titration and circular dichroism studies suggested that the old enzyme differs from the young even though the amino acid composition is probably unchanged and there is a lack of N or C terminal degradation during isolation.[62] The differences between young and old forms disappeared when the enzyme was denatured in 1.25 *M* guanidine and renatured.[63] After reformation, the young and old enolase activities were indistinguishable by thermal denaturation, protease inactivation, immunotitration with young antiserum, or circular dichroism. These results indicate that young and old enolases are conformational isomers.

Rates of incorporation of $^3$H-leucine into either total protein or enolase (precipitated using monospecific antibodies) showed large decreases between day 5 and day 32.[64] These decreases occurred whether synchrony was maintained by screening or by FUdR treatment. They were not due to altered uptake of radioactive leucine. Sharma et al.[64] conclude that, for both pure enolase and total protein, there is little change in the rate of uptake of label but a tenfold decrease in the rate of incorporation of label into protein and a threefold decrease in the rate of protein degradation. In summary, Rothstein and Sharma suggest that the lower specific activity of enolase seen in aged worms is due to a threefold decrease in turnover rate which leads to an accumulation of enzyme molecules which have undergone tertiary changes leading to an inactive form.

### F. Prolongation of Life by Caloric Restriction

When *C. elegans* is starved at the second larval molt, the animal undergoes an interruption in development and forms an alternative third larval stage referred to as dauer larvae.[29] This stage is especially resistant to dessication and other adverse environmental effects. Klass and Hirsh[11] found that even nematodes that had existed as dauers for as long as 60 days had normal adult lifespans (10 days under these conditions) after being fed and allowed to resume normal development. Thus, the dauer state appears to represent a "time out" from aging as well as from larval development.

Kline and Johnson[65] (Table 3), performed similar experiments on animals starved at the first larval stage, too early to allow dauer formation. This starvation almost immediately arrested development and resulted in prolongation of lifespan in direct proportion to the length of time starved, suggesting that starvation may stop the aging "clock" temporarily until development resumes. Similar results were obtained in animals arrested by starvation in other larval stages or as adults. However, starvation of adults causes cessation of egg laying, which unfortunately results in premature death of fecund hermaphrodites due to hatching of unlaid eggs within their bodies. Nevertheless, starvation of adults for as long as 2 days resulted in the prolongation of life in animals surviving this treatment.

Similar experiments by Mitchell et al.,[66] using axenic media to interrupt development of a *C. elegans* strain which was incapable of axenic growth (see Section III.H.3), showed a similar prolongation of total life (Table 4). Comparisons of these two sets of results suggest that complete starvation and treatment with axenic media are similar in their modes of action in prolonging life. The reader is also referred to Section III.B.4 for information on lifespan and food consumption.

## Table 3
## LIFESPANS AFTER STARVATION BLOCK[a]

| Length of block (days) | Lifespan (days) Mean ± SE[b] | Increase over control[c] (days) | Number of worms |
|---|---|---|---|
| 0[d] | 20.7 ± 0.90 | — | 40 |
| 2 | 21.5 ± 0.63 | 0.8 | 44 |
| 4 | 23.8 ± 0.78 | 3.1 | 19 |
| 6 | 25.4 ± 0.70 | 4.7 | 39 |
| 8 | 26.1 ± 1.13 | 5.4 | 29 |
| 10 | 32.6 ± 1.63 | 11.9 | 14 |

[a]   Unpublished data.[65]
[b]   Standard error.
[c]   Linear regression of increase in lifespan on length of block gives a slope of 1.07 ± 0.08.
[d]   Control.

## Table 4
## LIFE CYCLE CHARACTERISTICS OF ARRESTED N2 WORMS[a]

| Days arrested | Healthy adults (%) | Mean ± SE Juvenile period (days) | Adult period (days) | Total lifespan (days) | Fertility (progeny per healthy adult) |
|---|---|---|---|---|---|
| 0 (N = 35) | 100 | 2.5 ± 0.1 | 26 ± 2 | 29 ± 2 | 211 ± 9 |
| 15 (N = 34) | 72 | 17.5 ± 0.1 | 20 ± 1 | 37 ± 1 | — |
| 21 (N = 32) | 34 | 23.0 ± 0.1 | 17 ± 1 | 40 ± 1 | 165 ± 8 |
| 28 (N = 30) | 20 | 30.0 ± 0.1 | 20 ± 1 | 50 ± 1 | 99 ± 7 |

[a]   Unpublished data.[66]

## G. Drug Studies
### 1. DNA Inhibitors

FUdR has been widely used in synchronously senescing cultures as an inhibitor for blocking the production of viable progeny. FUdR (100 $\mu M$) blocks 96% of the total DNA synthesis in *T. aceti*.[67] Other DNA synthesis inhibitors that have been tested in *T. aceti* include 500 $\mu M$ hydroxyurea, which blocks 90% of total DNA synthesis, and 100 $\mu M$ aminopterin, which inhibits 92% of DNA synthesis.[67] DNA synthesis inhibition has not been directly tested in *C. elegans*, but FUdR has been found to cause sterility by blocking development of both eggs and larvae within 6 hr after addition in axenic culture (400 $\mu M$)[17] and in monoxenic culture (30 $\mu M$).[27] FUdR (400 $\mu M$, axenic media) inhibits the growth of *C. elegans* larvae less than 0.7 mm in length, whereas larger larvae are able to grow to lengths comparable to those of adults;[17] 30 $\mu M$ FUdR in monoxenic media permits near normal growth of first larval stage *C. elegans*. A length of 0.7 mm may correspond to the stage and time in larval development when all somatic mitoses have finished;[68] Kline and Johnson[65] observed that the onset of adult fertility is not blocked by starvation of larvae of this size or larger. Addition of FUdR (0.1 m$M$) to young *T. aceti* apparently does not block further development as monitored by average adult length of the animals.[67]

As suggested earlier, potential artifacts could be introduced both by indirect effects of the blockage of DNA synthesis and by secondary effects of the inhibitor on other processes. FUdR, even at concentrations as low as 25 $\mu M$, can cause abnormal vulval development in *C. elegans*, including extrusion of the gonad through the vulva.[26] Perhaps because vulval formation involves some of the last mitotic events to occur in larval development,[68] it may be a sensitive monitor of abnormal or blocked mitoses late in development. Addition of 25 $\mu M$ FUdR at 55 hr after hatch in axenic media induces complete sterility, while almost completely eliminating abnormal vulval and gonadal development.[26] Treatment with 30 $\mu M$ FUdR in monoxenic cultures at the first larval stage permits almost normal gonad development and normal numbers of oocytes and eggs.[27] A 400-$\mu M$ FUdR dose causes a 75% decrease in rate of movement (Figure 4),[14] but 25 (axenic) or 30 $\mu M$ (monoxenic) FUdR causes no alterations of this behavior.[26,27] However, even at the lower concentration, there is a decrease in the number of eggs laid in axenic cultures. Doses of FUdR as high as 50 m$M$ cause decreases in lifespan, whereas a lower dose (400 $\mu M$) causes an increase in lifespan from 21 to 29 days in axenic cultures and from 15 to 16 days in monoxenic cultures of *C. elegans*.[26] At still lower concentrations less than 10% increases in lifespan are observed.[26,27] We[69] have observed that addition of 400 $\mu M$ FUdR at 2 days prolongs life, while later addition does not. Several other age-related processes, such as the decrease in movement and pharyngeal pump rate, the accumulation of gut pigment, and gonadal atrophy and the resulting cessation of egg laying, proceed in the same manner with or without the addition of 25 $\mu M$ FUdR in axenic media[26] or 30 $\mu M$ FUdR in monoxenic media.[27]

In *T. aceti*, aminopterin (50 $\mu$g/m$\ell$), FUdR (100 $\mu$g/m$\ell$), and hydroxyurea (300 $\mu$g/m$\ell$) inhibit reproduction, while reproduction in *C. briggsae* is completely blocked by 50 $\mu$g/m$\ell$ of either aminopterin or FUdR.[70] As in *C. elegans*, these drugs also cause some vulval abnormalities.[70] Kisiel et al.[71] argue that aminopterin treatment of *C. briggsae* leads to an earlier onset of several characteristic features of senescence, noting an earlier increase in specific gravity, increased osmotic fragility, and fine structural changes in treated animals. They report mean lifespans of about 14 days and 18 days for *C. briggsae* treated with aminopterin and FUdR, respectively, as compared with untreated lifespans of 29 days;[70] for treated *T. aceti*, they report lifespans of about 15 days vs. control (untreated) lifespans of about 25 days. Gershon,[67] also using 100 $\mu$g/m$\ell$ of FUdR, saw no change in *T. aceti* lifespan as a result of treatment.

*2. Other Drugs*

Lifespan of *C. briggsae* has been prolonged by use of $\alpha$-tocopherol, a natural antioxidant, and $\alpha$-tocopherolquinone ($\alpha$TQ), a derivative.[72] Concentrations of 400 $\mu$g/m$\ell$ increased the average survival time from 35 $\pm$ 2 days to 46 $\pm$ 2 days. $\alpha$TQ had no effect on the reproductive period or on the total number of eggs laid. Addition of $\alpha$TQ at 1 day or at 10 days of age resulted in prolongation of life, while addition at 20 or 30 days had no effect. A noticeable increase in length of life was also apparent when the drug added at 1 day of age was removed at days 6 or 10, suggesting that the first 10 days of life are the most critical period for prolongation of life with this drug. Kisiel and Zuckerman[73] observed no effect of centrophenoxine on lifespan in *C. briggsae*, but did see a decrease in two markers normally associated with old age: worms treated with 17 m$M$ centrophenoxine had an average specific gravity of 1.09 at 21 days of age (as compared to 1.20 in untreated worms), and osmotic fragility decreased from 46 to 13%. A lower dose of centrophenoxine (6.8 m$M$) led to a 40% decrease in the volume of lipofuscin granules measured in cross sections of the worms. Similarly,[74] dimethylaminoethanol (DMAE) and p-chlorophenoxy acetic acid (PCA), two natural breakdown products of centrophenoxine, were found to decrease the volume of lipofuscin granules more than 10-fold in 21-day-old *C. briggsae*. Again, there was no effect on lifespan. Castillo et al.[75] saw little effect of either procaine hydrochloride or Gerovital H$_3$ on growth or development below 3.6 m$M$. Higher levels affected both growth and fecundity, as well as net negative charge on surface membranes as assayed by visualization with cationized ferritin. No effect on specific gravity was observed, but decreased osmotic fragility in 21-day-old worms led the investigators to be optimistic about possible life extension. No lifespan determinations were reported.

Following up an observation that cGMP accumulates in axenic cultures of aging *P. redivivus*, Willett et al.[56] added cGMP at concentrations of 50, 150, and 250 n*M* to axenic cultures of *C. elegans*. The higher two concentrations had significant effects on lifespan, increasing median length of life from 33 to 35 and 41 days, respectively.

## H. Genetic Studies

### 1. Dauer Larvae

The dauer state, which apparently is a "time-out" from normal aging (Section III.F), has also been found by Yeargers[31] to be highly resistant to the expected shortening of life after γ-irradiation. Even at doses up to 60 krads, there was no appreciable decrease in lifespan. This dosage is high enough to cause several recessive lethal events per genome. Unfortunately, no control was included in the experiment to determine the effects of this irradiation on the lifespan of normal third stage larvae or adult *C. elegans*. Klass[16] reported only a 37% shortening of life after exposure of young adults to 192,000 ergs/cm² of UV irradiation. In pilot experiments we saw no decrease in lifespan in either control third stage larvae or adults after similar levels of γ-irradiation.[76] Thus, the significance of Yeargers' report is in some doubt. Anderson[77] observed fourfold increases in the specific activity of superoxide dismutase in dauers as compared to normal larvae or adults.

### 2. Genetic Changes During Aging

Klass[16] showed that UV irradiation of animals at different ages reduces lifespan but has smaller effects on total lifespan reduction as the worms age. In a genetic recombination assay using two linkage group I markers, *dpy-5* and *unc-15*, Rose and Baillie[78] found that the recombination frequency of the last quartile of progeny was eight times lower than that of the first quartile.

### 3. A Mutation Conferring the Ability to Grow in Axenic Media

The Brenner wild-type strain (N2) of *C. elegans* shows very poor growth in axenic media. Mitchell[79] analyzed the genetic basis of this phenomenon by crossing a strain adapted for axenic growth (N2A) with males adapted for monoxenic growth and carrying the temperature-sensitive mutation, *b26*, which blocks sperm production at 25°C. All of the F1 males, but none of the F1 hermaphrodites, grew in axenic media, suggesting that there is a single recessive X-linked mutation responsible for this growth difference. Mitchell subsequently mapped this locus, *axe-1*, to the X chromosome 2.2% to the right of *dpy-3*. In an incomplete analysis of another spontaneous mutant strain (N2A2), Mitchell found that the growth of F1 progeny of a cross with N2 males was intermediate to that of the parental strains and that males were not appreciably different from hermaphrodites in growth ability. These findings suggest that the locus (or loci) responsible for axenic growth in this mutant is different from *axe-1* and probably is not sex-linked.

### 4. Lifespans of Genetically Marked Strains

In any genetic study one must be able to cross mutant strains to other stocks carrying genetic markers in order to perform mapping, complementation tests, suppression studies, etc. It is therefore necessary to have a series of good marker stocks (strains with obvious morphological, behavioral, or other types of mutations) which do not themselves show alterations in other characteristics, especially lifespan. With this end in mind we have begun a survey of temperature-sensitive *ts*, morphological, and behavioral mutants in *C. elegans*. Representative data are shown in Table 5. There are many ts developmental mutations which are without severe effects on lifespan at a temperature (20°C) intermediate to their nonpermissive temperature of 25.5°C and the permissive temperature of 16°C. Three behavioral mutations *(unc-2, unc-20, and unc-78)* also did not produce defects severe enough to in-

## Table 5
## LIFESPANS OF MUTANT STRAINS[a]

| Stock or allele | Description | Lifespan (days) Mean ± SE[b] | Number of worms scored | Probability of similarity to N2[c,d] |
|---|---|---|---|---|
| N2 | Wild-type | 17.7 ± 0.4 | 188 | — |
| b26 | Ts, fertilization | 17.1 ± 0.7 | 94 | 0.196 |
| b48 | | 15.4 ± 0.7 | 57 | 0.004 |
| b53 | | 17.9 ± 0.8 | 53 | 0.818 |
| b78 | | 17.1 ± 0.7 | 67 | 0.823 |
| b86 | | 17.4 ± 1.0 | 39 | 0.642 |
| b122 | | 16.8 ± 0.8 | 40 | 0.228 |
| b126 | | 16.2 ± 1.0 | 50 | 0.061 |
| b221 | | 16.6 ± 0.8 | 44 | 0.191 |
| b245 | | 14.4 ± 0.5 | 76 | <0.001 |
| b249 | | 16.3 ± 0.9 | 57 | 0.218 |
| b252 | Ts, fertilization and gonadogenesis | 11.8 ± 0.5 | 75 | <0.001 |
| N2 | Wild-type | 20.1 ± 0.8 | 31 | — |
| CB55 | Uncoordinated (unc-2) | 17.4 ± 0.5 | 42 | 0.07 |
| CB112 | Uncoordinated (unc-20) | 17.8 ± 0.6 | 47 | 0.85 |
| CB1217 | Uncoordinated (unc-78) | 17.8 ± 0.7 | 41 | 0.80 |
| CB130 | Dumpy (dpy-8) | 8.5 ± 0.6 | 26 | <0.001 |
| CB678 | Long (lon-2) | 16.9 ± 0.4 | 42 | <0.01 |
| N2 | Wild-type | 18.4 ± 0.4 | 81 | — |
| b26 | Ts, fertilization | 17.2 ± 0.7 | 45 | 0.218 |
| b245 | | 14.9 ± 0.6 | 34 | <0.001 |
| b252 | Ts, fertilization, and gonadogenesis | 11.4 ± 0.6 | 42 | <0.001 |
| CB1520 | XO hermaphrodites (her-1) | 17.6 ± 0.6 | 34 | 0.194 |
| CB3232 | XO hermaphrodites (her-1) | 16.2 ± 0.9 | 26 | 0.012 |
| N2 | Wild-type | 21.1 ± 0.7 | 39 | — |
| Bergerac | Ts, gonadogenesis | 20.0 ± 0.9 | 50 | 0.172 |
| b27 | | 13.9 ± 1.3 | 9 | 0.009 |
| b38 | | 14.9 ± 0.9 | 27 | 0.005 |
| b43 | | 17.7 ± 0.8 | 24 | 0.024 |
| b57 | | 17.5 ± 0.8 | 35 | 0.367 |
| b74 | | 18.1 ± 0.6 | 42 | 0.001 |
| b80 | | 14.9 ± 0.8 | 37 | <0.001 |
| N2 | Wild-type | 22.4 ± 0.6 | 41 | — |
| b41 | Ts, gonadogenesis | 19.2 ± 0.6 | 45 | 0.001 |
| b91 | | 18.5 ± 0.7 | 35 | <0.001 |
| b146 | | 21.4 ± 0.8 | 40 | 0.419 |
| b151 | | 14.4 ± 0.6 | 41 | <0.001 |
| b210 | Ts, gonadogenesis | 21.6 ± 1.0 | 40 | 0.822 |
| b221 | | 15.3 ± 0.5 | 50 | 0.001 |
| b441 | | 14.8 ± 0.8 | 32 | <0.001 |

[a]  Unpublished data.[30,65,69,76]

[b]  Standard error.

[c]  Probabilities less than 0.05 indicate significant lifespan differences.

[d]  Data are from five different experiments. Comparisons for each experiment are made for the N2 control run in the same experiment.

validate comparisons to the wild-type parental stocks. But two morphological mutants *(dpy-8* and *lon-2)*, which were short and squat *(dpy-8)* or longer than wild-type *(lon-2)*, both had lifespans in the same assay which were significantly shorter than wild-type. In yet another study, two *her-1* mutations which cause normally male XO worms to become fertile pseu-dohermaphrodites had no significant effect upon lifespan.

Lashlee and Johnson[69] asked if the decrease in lifespan in several ts mutants was the result of the mutation itself or of other mutations induced in the same mutagenesis and still present within the stock. If decreased lifespan resulted from other mutations, reisolates from a backcross of the mutant strain to wild-type should segregate stocks with lifespans more like those of the parental (N2) strain. For *b235*, *b244*, *b245*, and *b261* it has been possible to obtain reisolated mutant strains from the backcross which no longer have lifespans that differ significantly from that of the parental strain (Table 6). The results for one mutation *(b245)* are shown in more detail in Figure 7. For three of these four mutations, it appears that other genetic loci are responsible for shorter life and can be separated from the ts locus by genetic recombination. The data for *b244* are not as strong, since only marginally similar lifespans have been achieved.

If decreased lifespan is due to secondary mutations, one might predict that mutagenesis with ethyl methane sulfonate (EMS) following standard procedures[10] would yield stocks which have a detectable decrease in mean lifespan. This hypothesis was tested[80] by muta-genizing with EMS and then allowing the mutagenized animals to self-fertilize in the usual manner to produce an F1 population which was also allowed to self-fertilize and form an F2. As shown in Figure 8, a putative mutation would first be induced in the germ-line of the parental generation and would become heterozygous in the F1 and homozygous in some F2 individuals. Thus, if there are many recessive genes with sublethal effects on lifespan, we would expect that there should be a reduced lifespan in the F2 but not in earlier generations. The data (Table 7) support this notion. No decrease in lifespan was observed in either the treated worms or their F1 progeny when compared to untreated controls. However, there was a significant decrease in mean lifespan of the F2 animals, each of which was homozygous for 20 or more mutational alterations as a result of the EMS treatment.[10]

Finally, we examined two mutations, *unc-15* and *unc-54*, which have altered paramyosin and myosin, respectively.[81] Both of these mutant strains have severely shortened lifespans (Table 8), and both show severe defects in their ability to lay eggs as a secondary effect of their mutant vulval musculature. If lifespan shortening is due primarily to this secondary effect in that animals tend to die from internal hatch or suffer severe problems as a result of the failure to lay eggs, we should be able to suppress life shortening by blocking viable progeny formation. Double mutants of *unc-15* or *unc-54* and *b26* or *b245* were constructed and grown at 25°C to block progeny production. This procedure restored normal lifespans in *unc-15*, *b26* and *unc-15*, *b245* double mutants and near normal lifespan in *unc-54*, *b26* double mutants. This genetic suppression of the lifespan defect is direct evidence for a role of egg production in shortening lifespan in these backgrounds.

## 5. Isolation of Newly Induced Longevity Mutants

Arguing that shorter lifespan can be viewed as a by-product of any defect in the metabolic machinery, Klass[16,25a] screened for more specific mutations by looking for EMS-induced alterations which generated stocks with *longer* lifespans (Figure 8). He took advantage of the self-fertilizing hermaphroditic mode of reproduction of *C. elegans* in screening 8000 F2 clones resulting from two rounds of self-fertilization of EMS-treated hermaphrodites. Using the *b26* mutation to block sperm formation in the aging clones, he isolated eight strains that lived at least 20% longer than wild-type. Further analysis of these strains showed that two entered the dauer state even when fed, so that their longer lifespan was presumably due to the absence of aging in dauers. One strain was chemotaxis-defective, and the others showed alterations in pharyngeal pump rate, which is a measure of the rate of food ingestion. When the relationship between the pump rate and the increase in lifespan was examined, a good correlation was found (Figure 9). This result suggests that the major reason for the longer

**Table 6**
**LIFESPAN OF STOCKS REISOLATED FROM BACKCROSSES TO N2[a]**

| Stock or allele | Isolate number | Lifespan (days) Mean ± SE[b] | Number of worms | Probability of similarity to N2 |
|---|---|---|---|---|
| N2 | | 21.6 ± 0.9 | 43 | — |
| b235 | | 19.5 ± 1.1 | 38 | 0.18 |
| b235 | 4 | 15.9 ± 1.0 | 35 | <0.001 |
| b235 | 7 | 23.2 ± 0.9 | 46 | 0.28 |
| b235 | 8 | 21.5 ± 0.9 | 49 | 0.83 |
| b235 | 9 | 19.0 ± 1.3 | 40 | 0.10 |
| b244 | — | 18.2 ± 0.8 | 39 | 0.01 |
| b244 | 1 | 19.2 ± 0.9 | 46 | 0.05 |
| b244 | 2 | 15.9 ± 1.0 | 39 | <0.001 |
| b244 | 4 | 16.7 ± 0.9 | 43 | <0.001 |
| b244 | 5 | 19.4 ± 1.0 | 38 | 0.15 |
| b261 | — | 22.0 ± 0.7 | 38 | 0.77 |
| b261 | 2 | 22.4 ± 0.9 | 41 | 0.60 |
| b261 | 6 | 22.0 ± 1.2 | 35 | 0.86 |
| N2 | — | 18.3 ± 0.8 | 39 | — |
| b245 | — | 12.9 ± 0.8 | 26 | <0.001 |
| b245 | 1 | 14.7 ± 0.7 | 33 | 0.001 |
| b245 | 3 | 14.8 ± 0.8 | 27 | 0.004 |
| b245 | 5 | 14.4 ± 0.7 | 30 | 0.001 |
| b245 | 12 | 15.7 ± 0.8 | 31 | 0.058 |
| b245 | 21 | 18.5 ± 0.6 | 37 | 0.557 |
| b245 | 23 | 15.2 ± 0.7 | 34 | 0.009 |

[a]   Unpublished data.[69]
[b]   Standard error.

lifespans of these six mutant strains may have been a lower rate of food ingestion, which has previously been shown to result in prolongation of life (see Section III.D.2).

Klass[16] interprets these data as indicating that there are very few (perhaps no more than one) specific "aging" genes, defined as genes which are activated at maturity and which lead to subsequent senescence. I find it very interesting that single-gene mutations can produce longer lived strains, and we have begun a genetic analysis of these strains to determine the total number of genes involved, their allelism, and their location.

*6. Lack of Heterosis and Inbreeding Depression for Lifespan*

Crosses between Bristol and Bergerac strains, as well as crosses between Bristol and a number of other wild-type strains, have yielded a very surprising result.[12,13] In contrast to other animal species where the F1 hybrid progeny of a cross between two inbred strains exhibit lifespans that are much longer than those of either parental strain (heterosis),[82] the lifespans of a number of hybrid F1 populations have been found to be similar to or shorter than the lifespans of their respective parental strains[83] (Table 9). An accepted explanation for the heterosis effect in sexual species is that wild populations of sexually reproducing species are heterozygous at many loci and thus have optimized their fitness (including lifespan) under these conditions. When these types of species are maintained in a laboratory environment, the stocks are either intentionally or unintentionally inbred, which results in a loss of the heterozygous condition at many loci and a consequent drop in fitness leading to decreased lifespans (inbreeding depression). When these stocks are crossed to other stocks maintained separately and therefore inbred for other alleles, the normal heterozygous situation

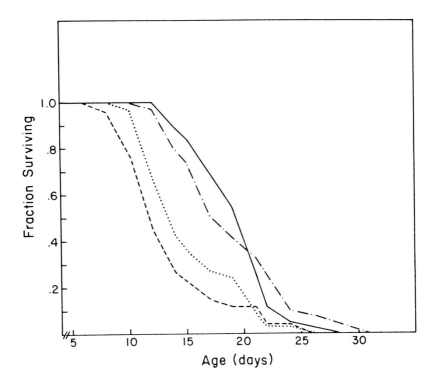

FIGURE 7.  Survival curves for N2 (–•–); DH245 (-----), a Ts mutant strain with defective spermatogenesis; and two Ts F2 reisolates from a backcross of DH245 to N2: isolate #1 (······), isolate #2 (————). For mean lifespans and other information, see Table 6.[69]

is regained and lifespan is extended. In contrast, in an inbreeding hermaphroditic organism such as *C. elegans*, which must optimize its fitness in the absence of significant heterozygosity in wild populations, we might expect to observe neither inbreeding depression of lifespan in the laboratory nor lengthening of lifespan in interstrain crosses. This single fact about the life history of *C. elegans* may make it the only organism in which quantitative genetic analyses of lifespan can be simply carried out with easily interpretable results.

Lifespan in *C. elegans* is a quantitative trait which can be examined in crosses between inbred strains of different genetic backgrounds without worrying about inbreeding depression and subsequent problems with potential loss of phenotype. Indeed, the results of the analysis of recombinant inbred strains (see Section III.G.7) are consistent with the absence of inbreeding depression in that lifespans of these strains appear to have diverged from the lifespans of the original parental strains (19 days) more or less equally in both directions (Figure 10).

## 7. Genetic Variability in Lifespan

Johnson and Wood[13] have estimated the amount of genetic variability in lifespan using two wild-type strains of *C. elegans*, var. Bristol (or N2) and var. Bergerac. An analysis of the increase in the variance of length of life in the F2 derived from crosses between these strains showed highly significant increases in variability leading to heritability estimates of about 40%. This result was confirmed by isolating individual F2 animals and allowing them to self-fertilize for 18 additional generations, forming a series of recombinant inbred strains. The mean lifespan of these recombinant inbreds was regressed onto the lifespan of the F2 parents to obtain another estimate of the amount of genetic variability in lifespan.[13] This procedure resulted in values of 51 and 43% in two independent experiments. A third method

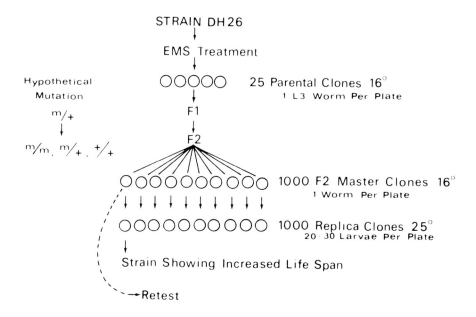

FIGURE 8.    Mutagenesis was carried out by the procedure of Brenner[10] using ethyl methane sulfonate (EMS). L3 stage larvae were transferred onto NGM agar plates seeded with *E. coli* strain OP50 after EMS treatment. These "parental clones" were allowed to undergo self-fertilization for two generations at permissive temperature (16°C) after which 1000 F2 progeny were picked and individually transferred to fresh NGM plates (one worm per plate) at 16°C. These "master clones" were allowed to reproduce for several generations after which samples of 20 to 30 first-stage larvae were removed from each clone and transferred to individual plates seeded with a lawn of *E. coli* at restrictive temperature (25°C). These "replica clones" were allowed to mature, age, and die; during this process, they were checked at 10, 20, 30, 40, 60, and 70 days after transfer to 25°C to determine the presence of live worms. The mean lifespan at 25°C is 8.9 ± 1.0 days and the maximum lifespan is 12 days for the wild-type.[16] Those clones having live worms after 20 days were retested by retrieving the master clone and measuring the lifespan at 25°C. Because of the increase in lifespan caused by dietary restriction,[11,16] care must be taken to insure that the worms on the replica clones have a greater than adequate food supply.[25]

**Table 7**
**LIFESPAN AFTER EMS TREATMENT[a]**

| Stock | Number of worms | Lifespan (days) Mean ± SE[b] | Probability of similarity to N2 |
|---|---|---|---|
| N2 control | 40 | 20.1 ± 0.7 | — |
| N2 EMS treated | 40 | 20.2 ± 0.5 | 0.98 |
| F1 of EMS treated | 137 | 19.9 ± 0.4 | 0.82 |
| F2 of EMS treated | 121 | 16.9 ± 0.3 | 0.001 |

[a]    Unpublished data.[80]
[b]    Standard error.

of estimating heritability of lifespan is to compare variability within recombinant inbred strains to variability between strains. This procedure gave a heritability value of 19%.[13] Thus, 20 to 50% of the variation in length of life in this population seems to be due to the combined effects of genes that have small individual effects on lifespan.

The recombinant inbred strains have lifespans that range from 10 to 31 days (Figure 10). Thus, the recombinant inbreds are themselves a rich source of genetic variation in length

## Table 8
## LIFESPANS AND SUPPRESSION IN *unc-15* AND *unc-54* STOCKS[a]

| Genotype | Number of worms | Lifespan (days) Mean ± SE[b] | Probability of similarity to N2[c] |
|---|---|---|---|
| N2 | 46 | 15.5 ± 0.7 | |
| *unc-54*[d] | 50 | — | — |
| *unc-15* | 20 | 8.6 ± 0.7 | <0.001 |
| b245 | 48 | 15.2 ± 0.3 | 0.810 |
| b26 | 49 | 14.5 ± 0.4 | 0.061 |
| *unc-54*, b26 | 39 | 13.1 ± 0.7 | 0.013 |
| *unc-15*, b245 | 32 | 15.6 ± 0.7 | 0.964 |
| *unc-15*, b26 | 46 | 14.9 ± 0.7 | 0.733 |

[a] Unpublished data.[80]
[b] Standard error.
[c] Probabilities less than 0.05 indicate significant lifespan differences.
[d] All animals died as "bags of worms" from internal egg hatching.

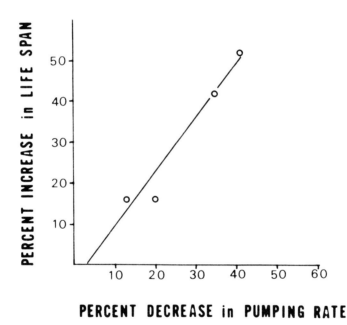

FIGURE 9. Linear regression of percent increase in lifespan on percent decrease in pharyngeal pumping rate.[25]

of life. They show more than a 12-fold extension in the range of mean lifespans over that displayed by the parental strains.[13] These results taken as a whole suggest that there are large amounts of genetic control at many loci and that we should be able to extend lifespan of the worm dramatically in further selection experiments. Gompertz plots (see Section III.B.2) for the parental strains and for three of the recombinant inbred strains are presented in Figure 11, where the exponential increase with age in the hazard rate for each strain has been linearized by plotting on a semilogarithmic scale. The linear regression lines resulting from the best fit to these data show that the three recombinant inbred strains have the same initial susceptibility to aging but age at quite different rates. More detailed investigation will be required to evaluate the significance of this finding.

**Table 9**
## LIFESPANS OF F1 HYBRIDS OF CROSSES BETWEEN WILD-TYPE STRAINS[a]

| Strains | Sex[b] | Lifespan (days) Mean ± SE[c] | Number of worms | Probability value[d] compared to DH26 | Wild-type strains |
|---|---|---|---|---|---|
| Parental stocks | | | | | |
| DH26 | H | 16.1 ± 0.6 | 81 | | |
| N2 | H | 20.3 ± 0.5 | 159 | | |
| N2 | M | 13.8 ± 0.3 | 175 | | |
| PA-1 | H | 18.1 ± 1.6 | 43 | | |
| PA-1 | M | 15.4 ± 0.5 | 89 | | |
| Pac-1 | H | 18.4 ± 0.9 | 18 | | |
| Pac-1 | M | 13.4 ± 0.3 | 46 | | |
| GA-1 | H | 16.4 ± 0.5 | 69 | | |
| GA-1 | M | 12.8 ± 0.5 | 69 | | |
| CL2A | H | 18.5 ± 0.8 | 51 | | |
| CL2A | M | 21.9 ± 0.6 | 93 | | |
| F1 hybrids | | | | | |
| N2/DH26 | H | 18.0 ± 0.6 | 75 | 0.05 | 0.02 |
| N2/DH26 | M | 12.5 ± 0.3 | 136 | | 0.02 |
| PA-1/DH26 | H | 22.5 ± 1.3 | 21 | 0.00 | 0.03 |
| PA-1/DH26 | M | 14.2 ± 0.5 | 90 | | 0.123 |
| Pac-1/DH26 | H | 16.5 ± 0.6 | 53 | 0.56 | 0.07 |
| Pac-1/DH26 | M | 13.9 ± 0.4 | 102 | | 0.612 |
| GA-1/DH26 | H | 16.5 ± 0.8 | 58 | 0.70 | 0.915 |
| GA-1/DH26 | M | 12.6 ± 0.3 | 98 | | 0.974 |
| CL2A/DH26 | H | 19.6 ± 1.0 | 36 | 0.003 | 0.504 |
| CL2A/DH26 | M | 18.1 ± 0.6 | 90 | | 0.000 |

[a]  Unpublished data.[82]
[b]  H = hermaphrodite; M = male.
[c]  Standard error.
[d]  Probability values obtained by the Gehan statistic. Values less than 0.05 indicate significant lifespan differences.

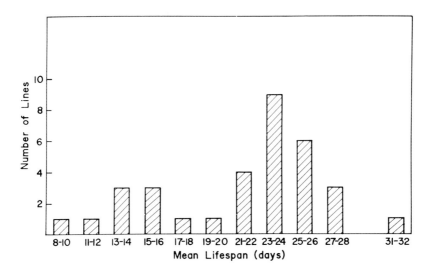

FIGURE 10.    Histogram showing the distribution of mean lifespan for 33 recombinant inbred strains[12] derived as described in the text and by Johnson and Wood.[13]

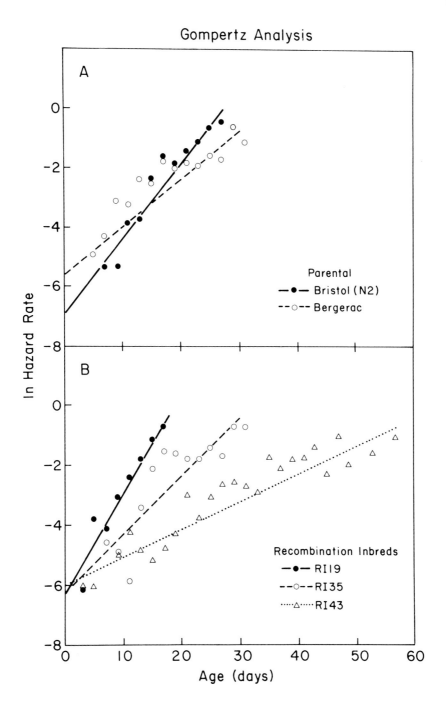

FIGURE 11. Hazard rate (age-specific death rate) calculated as described in the caption for Figure 2. (A) Parental strains; (B) three recombinant inbred strains derived from crosses between the two parental strains.[13,76]

## IV. SUMMARY

There is no doubt that *C.elegans* is remarkably well suited for use as an experimental organism in studies of the biological basis of aging and as an animal model for studying factors that may affect length of life. Among other reasons, its short (19-day) average lifespan, the ease of analysis, and its unique mode of reproduction by self-fertilization (which may be responsible for the observed absence of inbreeding depression or heterosis) make this an organism of choice for studies of aging. In particular, the development of genetic stocks with large increases in mean lifespan promises to be a fruitful approach to identifying the molecular products responsible for longer life.

Several age-related phenomena which have so far been observed only in other species of nematodes should be studied in *C. elegans*. This is particularly true of morphological changes. Much more work is also needed at both the molecular and the genetic level.

This reviewer is very optimistic that exciting advances in our understanding of aging will result from the use of *C. elegans* in genetic and molecular analyses. Furthermore, its use in pharmacological studies of agents that may extend lifespan will contribute greatly to our knowledge of metazoan aging and perhaps even to our ability to intervene in the process of senescence.

## ACKNOWLEDGMENTS

I am indebted to Rebecca G. Miles for her expert and efficient help in maintaining coherence in this manuscript and to Barb Kirwin for rapid translation of somewhat disorganized notes. This work was supported in part by an award (AG-01236) from the National Institute on Aging and by BRSG grant RR-07013-16 awarded to the University of Colorado by the Biomedical Research Support Grant Program, Division of Research Resources, National Institutes of Health.

## REFERENCES

1. **Zuckerman, B. M., Ed.,** *Nematodes as Biological Models,* Vols. 1, 2, Academic Press, N.Y., 1980.
2. **Riddle, D. L.,** Developmental biology of *Caenorhabditis elegans:* symposium introduction, *J. Nematol.,* 14, 238, 1982.
3. **Horvitz, H. R. and Sternberg, P. W.,** Nematode postembryonic cell lineages, *J. Nematol.,* 14, 240, 1982.
4. **Edgar, R. S., Cox, G. N., Kusch, M., and Politz, J. C.,** The cuticle of *Caenorhabditis elegans, J. Nematol.,* 14, 248, 1982.
5. **Ward, S., Roberts, T. M., Nelson, G. A., and Argon, Y.,** The development and motility of *Caenorhabditis elegans* spermatozoa, *J. Nematol.,* 14, 259, 1982.
6. **Wood, W. B., Laufer, J. S., and Strome, S.,** Developmental determinants in embryos of *Caenorhabditis elegans, J. Nematol.,* 14, 267, 1982.
7. **Johnson, T. E.,** Aging in *Caenorhabditis elegans,* in *Review of Biological Research in Aging,* Rothstein, M., Ed., Alan R. Liss, N.Y., 1983, 37.
8. **Rothstein, M.,** Effects of aging on enzymes, in *Nematodes as Biological Models,* Vol. 2, Zuckerman, B. M., Ed., Academic Press, N.Y., 1980, 29.
9. **Zuckerman, B. M. and Himmelhoch, S.,** Nematodes as models to study aging, in *Nematodes as Biological Models,* Vol. 2, Zuckerman, B. M., Ed., Academic Press, N.Y., 1980, 4.
10. **Brenner, S.,** The genetics of *Caenorhabditis elegans, Genetics,* 77, 71, 1974.
11. **Klass, M. and Hirsh, D.,** Nonaging developmental variant of *Caenorhabditis elegans, Nature (London),* 260, 523, 1976.

12. **Johnson, T. E.,** Unpublished observations, 1982.

13. **Johnson, T. E. and Wood, W. B.,** Genetic analysis of lifespan in *Caenorhabditis elegans, Proc. Natl. Acad. Sci. U.S.A.,* 79, 6603, 1982.

14. **Bolanowski, M. A., Russell, R. L., and Jacobson, L. A.,** Quantitative measures of aging in the nematode *Caenorhabditis elegans.* I. Population and longitudinal studies of two behavioral parameters, *Mech. Aging Dev.,* 15, 279, 1981.

15. **Emmons, S. W., Klass, M. R., and Hirsh, D.,** Analysis of the constancy of DNA sequences during development and evolution of the nematode *Caenorhabditis elegans, Proc. Natl. Acad. Sci. U.S.A.,* 76, 1333, 1979.

16. **Klass, M. R.,** Aging in the nematode *Caenorhabditis elegans:* major biological and environmental factors influencing life span, *Mech. Aging Dev.,* 6, 413, 1977.

17. **Mitchell, D. H., Stiles, J. W., Santelli, J., and Sanadi, D. R.,** Synchronous growth and aging of *Caenorhabditis elegans* in the presence of fluorodeoxyuridine, *J. Gerontol.,* 34, 28, 1979.

18. **Rothstein, M. and Sharma, H. K.,** Altered enzymes in the free-living nematode, *Turbatrix aceti,* aged in the absence of fluorodeoxyuridine, *Mech. Aging Dev.,* 8, 175, 1978.

19. **Tilby, M. J. and Moses, V.,** Nematode aging. Automatic maintenance of age synchrony without inhibitors, *Exp. Gerontol.,* 10, 213, 1975.

20. **Croll, N. A., Smith, J. M., and Zuckerman, B. M.,** The aging process of the nematode *Caenorhabditis elegans* in bacterial and axenic culture, *Exp. Aging Res.,* 3, 175, 1977.

21. **Russell, R. L.,** Personal communication, 1982.

22. **Johnson, T. E., Lashlee, C. H., and McCaffrey, G.,** Quantitative genetics of lifespan in *Caenorhabditis elegans, Genetics,* 97, S53, 1981.

23. **Johnson, T. E., McCaffrey, G., and Lashlee, C. H.,** Genetic control of aging in *Caenorhabditis elegans, Genetics,* 96, S51, 1980.

24. **Klass, M. R. and Hirsh, D.,** Sperm isolation and biochemical analysis of the major sperm protein from *Caenorhabditis elegans, Dev. Biol.,* 84, 299, 1981.

25. **Klass, M. R., Nguyen, P. N., and De Chavigny, A.,** Age-correlated changes in the DNA template in the nematode *Caenorhabditis elegans, Mech. Aging Dev.,* 22, 253, 1983.

25a. **Klass, M. R.,** A method for the isolation of longevity mutants in the nematode *Caenorhabditis elegans, Mech. Aging Dev.,* 22, 279, 1983.

25b. **Klass, M. R.,** Personal communication, 1982.

26. **Gandhi, S., Santelli, J., Mitchell, D. H., Stiles, J. W., and Sanadi, D. R.,** A simple method for maintaining large, aging populations of *Caenorhabditis elegans, Mech. Aging Dev.,* 12, 137, 1980.

27. **Mitchell, D. H. and Santelli, J.,** Fluorodeoxyuridine as a reproductive inhibitor in *Caenorhabditis elegans:* applications to aging research, manuscript submitted, 1982.

28. **Herman, R. K. and Horvitz, H. R.,** Genetic analysis of *Caenorhabditis elegans,* in *Nematodes as Biological Models,* Vol. 1, Zuckerman, B. M., Ed., Academic Press, N.Y., 1980, 227.

29. **Riddle, D. L.,** Developmental genetics of *Caenorhabditis elegans,* in *Nematodes as Biological Models,* Vol. 1, Zuckerman, B. M., Ed., Academic Press, N.Y., 1980, 263.

30. **McCaffrey, G. and Johnson, T. E.,** Unpublished data, 1981.

31. **Yeargers, E.,** Effect of γ-radiation on dauer larvae of *Caenorhabditis elegans, J. Nematol.,* 13, 235, 1981.

32. **Gehan, E. A.,** A generalized Wilcoxon test for comparing arbitrarily singly-censored samples, *Biometrika,* 52, 203, 1965.

33. **Hull, C. H. and Nie, N. H., Eds.,** SPSS update: new procedures and facilities for releases 7 and 8, McGraw-Hill, N.Y., 1979.

34. **Peto, R., Pike, M. C., Armitage, P., Breslow, N. E., Cox, D. R., Howard, S. V., Mantel, N., McPherson, K., Peto, J., and Smith, P. G.,** Design and analysis of randomized clinical trials requiring prolonged observation of each patient. II. Analysis and examples, *Br. J. Cancer,* 35, 1, 1977.

35. **Miller, R. G., Jr.,** *Survival Analysis,* John Wiley & Sons, N.Y., 1981.

36. **Raines, M. and Johnson, T. E.,** Unpublished program, 1980.

37. **Sacher, G. A.,** Life table modification and life prolongation, in *Handbook of the Biology of Aging,* Finch, C. E. and Hayflick, L., Eds., Von Nostrand Reinhold, N.Y., 1977, 582.

38. **Zuckerman, B. M., Himmelhoch, S., Nelson, B., Epstein, J., and Kisiel, M.,** Aging in *Caenorhabditis briggsae, Nematologica,* 17, 478, 1971.

39. **Byerly, L., Cassada, L., and Russell, R. L.,** The life cycle of the nematode *Caenorhabditis elegans.* I. Wild-type growth and reproduction, *Dev. Biol.,* 51, 23, 1976.

40. **Beguet, B.,** The persistence of processes regulating the level of reproduction in the hermaphrodite nematode, *Caenorhabditis elegans,* despite the influence of parental aging, over several consecutive generations, *Exp. Gerontol.,* 7, 207, 1972.

41. **Beguet, B. and Brun, J. L.,** Influence of parental aging on the reproduction of the F1 generation in a hermaphrodite nematode *Caenorhabditis elegans, Exp. Gerontol.,* 7, 196, 1972.

42. **Epstein, J., Himmelhoch, S., and Gershon, D.,** Studies on aging in nematodes. III. Electronmicroscopical studies on age-associated cellular damage, *Mech. Aging Dev.,* 1, 245, 1972.
43. **Kisiel, M. J., Castillo, J. M., Zuckerman, L. S., Zuckerman, B. M., and Himmelhoch, S.,** Studies on aging *Turbatrix aceti, Mech. Aging Dev.,* 4, 81, 1975.
44. **Zuckerman, B. M., Himmelhoch, S., and Kisiel, M.,** Fine structure changes in the cuticle of adult *Caenorhabditis briggsae* with age, *Nematologica,* 19, 109, 1973.
45. **Hogger, C. H., Estey, R. H., Kisiel, M. J., and Zuckerman, B. M.,** Surface scanning observations of changes in *Caenorhabditis briggsae* during aging, *Nematologica,* 23, 213, 1977.
46. **Himmelhoch, S., Kisiel, M. J., and Zuckerman, B. M.,** *Caenorhabditis briggsae:* electron microscope analysis of changes in negative surface charge density of the outer cuticular membrane, *Exp. Parasitol.,* 41, 118, 1977.
47. **Croll, N. A.,** Components and patterns in the behavior of the nematode *Caenorhabditis elegans, J. Zool.,* 176, 159, 1975.
48. **Dusenbery, D. B.,** Behavior of free-living nematodes, in *Nematodes as Biological Models,* Vol. 1, Zuckerman, B. M., Ed., Academic Press, N.Y., 1980, 127.
49. **Hosono, R.,** Age dependent changes in the behavior of *Caenorhabditis elegans* on attraction to *Escherichia coli, Exp. Gerontol.,* 13, 31, 1978.
50. **Epstein, H. F., Isachsen, M. M., and Suddleson, E. A.,** Kinetics of movement of normal and mutant nematodes, *J. Comp. Physiol.,* 110, 317, 1976.
51. **Croll, N. A.,** Indolealkylamines in the coordination of nematode behavioral activities, *Can. J. Zool.,* 53, 894, 1975.
52. **Kisiel, M. J. and Zuckerman, B. M.,** Studies on aging of *Turbatrix aceti, Nematologica,* 20, 277, 1975.
53. **Hosono, R., Sato, Y., Aizawa, S. I., and Mitsui, Y.,** Age-dependent changes in mobility and separation of the nematode *Caenorhabditis elegans, Exp. Gerontol.,* 15, 285, 1980.
54. **Erlanger, M. and Gershon, D.,** Studies on aging in nematodes. II. Studies of the activities of several enzymes as a function of age, *Exp. Gerontol.,* 5, 13, 1970.
55. **Bolanowski, M. A., Jacobson, L. A., and Russell, R. L.,** Quantitative measures of aging in the nematode *Caenorhabditis elegans.* II. Lysosomal hydrolases as markers of senescence, *Mech. Aging Dev.,* 21, 295, 1983.
56. **Willett, J. D., Rahim, I., Geist, M., and Zuckerman, B. M.,** Cyclic nucleotide exudation by nematodes and the effects on nematode growth, development and longevity, *Age,* 3, 82, 1980.
57. **Link, C., Russell, R. L., and Jacobson, L.,** Personal communication, 1981.
58. **Zuckerman, B. M., Nelson, B., and Kisiel, M.,** Specific gravity increase of *Caenorhabditis briggsae* with age, *J. Nematol.,* 4, 261, 1972.
59. **Searcy, D. G., Kisiel, M. J., and Zuckerman, B. M.,** Age related increase of cuticle permeability in the nematode *Caenorhabditis briggsae, Exp. Aging Res.,* 2, 293, 1976.
60. **Rothstein, M.,** The formation of altered enzymes in aging animals, *Mech. Aging Dev.,* 9, 197, 1979.
61. **Sharma, H. K. and Rothstein, M.,** Serological evidence for the alteration of enolase during aging, *Mech. Aging Dev.,* 8, 341, 1978.
62. **Sharma, H. K. and Rothstein, M.,** Age-related changes in the properties of enolase from *Turbatrix aceti, Biochemistry,* 17, 2869, 1978.
63. **Sharma, H. K. and Rothstein, M.,** Altered enolase in aged *Turbatrix aceti* results from conformational changes in the enzyme, *Proc. Natl. Acad. Sci. U.S.A.,* 77, 5865, 1980.
64. **Sharma, H. K., Prasanna, H. R., Lane, R. S., and Rothstein, M.,** The effects of age on enolase turnover in the free-living nematode, *Turbatrix aceti, Arch. Biochem. Biophys.,* 194, 275, 1979.
65. **Kline, S. and Johnson T. E.,** Unpublished observations, 1981.
66. **Mitchell, D., Kemal, R., and Foy, J.,** Personal communication, 1981.
67. **Gershon, D.,** Studies on aging in nematodes. I. The nematode as a model organism for aging research, *Exp. Gerontol.,* 5, 7, 1970.
68. **Sulston, J. E. and Horvitz, H. R.,** Post-embryonic cell lineages of the nematode, *Caenorhabditis elegans, Dev. Biol.,* 56, 110, 1977.
69. **Lashlee, C. H. and Johnson, T. E.,** Unpublished observations, 1981.
70. **Kisiel, M., Nelson, B., and Zuckerman, B. M.,** Effects of DNA synthesis inhibitors on *Caenorhabditis briggsae and Turbatrix aceti, Nematologica,* 18, 373, 1972.
71. **Kisiel, M. J., Himmelhoch, S., and Zuckerman, B. M.,** *Caenorhabditis briggsae:* effects of aminopterin, *Exp. Parasitol.,* 36, 430, 1974.
72. **Epstein, J. and Gershon, D.,** Studies on aging in nematodes. IV. The effect of antioxidants on cellular damage and life span, *Mech. Aging Dev.,* 1, 257, 1972.
73. **Kisiel, M. J. and Zuckerman, B. M.,** Effects of centrophenoxine on the nematode *Caenorhabditis briggsae, Age,* 1, 17, 1978.

74. **Zuckerman, B. M. and Barrett, K. A.,** Effects of PCA and DMAE on the nematode *Caenorhabditis briggsae, Exp. Aging Res.,* 4, 133, 1978.
75. **Castillo, J. M., Kisiel, M. J., and Zuckerman, B. M.,** Studies on the effects of two procaine preparations on *Caenorhabditis briggsae, Nematologica,* 21, 401, 1975.
76. **Robinson, L. A. and Johnson, T. E.,** Unpublished data, 1982.
77. **Anderson, G. L.,** Superoxide dismutase activity in dauer larvae of *Caenorhabditis elegans* (Nematodai Rhabditidae), *Can. J. Zool.,* 60, 288, 1982.
78. **Rose, A. M. and Baillie, D. L.,** Effect of temperature and parental age on recombination and nondisjunction in *Caenorhabditis elegans, Genetics,* 92, 409, 1979.
79. **Mitchell, D. H.,** manuscript in preparation, 1982.
80. **Johnson, T. E., McCaffrey, G., and Lashlee, C. H.,** Unpublished data, 1980.
81. **Zengel, J. M. and Epstein, H. F.,** Muscle development in *Caenorhabditis elegans:* a molecular genetic approach, in *Nematodes as Biological Models,* Vol. 1, Zuckerman, B. M., Ed., Academic Press, N.Y., 1980, 74.
82. **Comfort, A.,** *Aging: The Biology of Senescence,* Holt, Rinehart, & Winston, N.Y., 1964, 173.
83. **Chu, A., Johnson, K., Lashlee, C. H., and Johnson, T. E.,** Unpublished data, 1980.

Chapter 5

# DEVELOPMENT AND FUNCTION OF THE LEECH NERVOUS SYSTEM

**William B. Kristan, Jr., David A. Weisblat, and Tricia Radojcic**

## TABLE OF CONTENTS

# I. INTRODUCTION

Invertebrate nervous systems offer many advantages over mammalian or other vertebrate nervous systems for sorting out the causes and specific mechanisms of aging. For instance, many invertebrates have a relatively short lifespan and a large number of individuals can be maintained rather easily. More importantly, individual neurons can be identified and studied morphologically, physiologically, and biochemically in many invertebrates. In particular, the nervous system of the leech has several advantages over that of other invertebrates as well as vertebrates:

1. Much of the nervous system has a repeated modular organization, so that there are multiple copies of homologous neuron clusters within an individual.
2. The nervous system consists of a small number of neurons (about 200 in each module), about 25% of which have been identified uniquely.
3. The neurons are large enough in many species to allow convenient electrophysiological recordings.
4. The nervous system is quite accessible. The whole nervous system can be dissected as a unit very quickly. It can be pinned out conveniently, due to a protective sheath that gives it great mechanical strength. Single neuronal somata can be viewed clearly using simple darkfield illumination through a dissecting microscope.

To date, aging has been characterized much more extensively in vertebrate nervous systems than in invertebrates. Functional decline has been characterized in sensory and motor systems, in reflex responses, and in learning and memory capabilities.[17,121] In the extreme condition called senile dementia, this decline is associated with loss of neurons (either grossly[11,72] or regionally[16,95]), cytological changes, such as neurofibrilary tangles, senile plaques, autophagic vacuoles and lipofuscin masses,[11,18,42,92,105,146] and impairment of chemical synaptic transmission,[26,64,103,105] particularly in cholinergic synapses.[21] However, the primary cause has not been pinpointed for any of these organic changes, nor have any of them been shown to cause dementia.

Recently, electrophysiological investigations of single neurons have been undertaken in an effort to define more precisely the effects of aging on the brain. Both hippocampal neurons[63,64] and cerebellar Purkinje cells[101,102] show electrophysiological deficits in aged animals, along with degenerative changes and lipofuscin accumulation. However, the correlation between physiological and morphological deficits is not always present. For instance, extensive lipofuscin accumulation in olivary nuclear neurons does not appear to affect the electrophysiological functioning of their climbing fiber afferents to the Purkinje cells.[102] Also, despite the electrophysiological and morphological changes in rat hippocampal neurons, they continue to take up deoxyglucose normally.[62,113]

Even this cursory review points out the frustrating difficulties inherent in establishing causality in the vertebrate, particularly the mammalian, brain. There are simply so many nerve cells and so many parts of the brain involved in any function that it is nearly impossible to establish more than correlations between neuronal and behavioral changes. Simpler, invertebrate nervous systems have yet to be studied systematically for this purpose. Preliminary studies have shown anatomical[80,115] and physiological[96,97] changes in neurons of older individuals of some invertebrate species. To date, there have been no studies of the effects of aging on the leech nervous system. However, given its simplicity, accessibility, and widespread use,[86] the leech nervous system could readily be used to study the effects of aging. In addition, this nervous system could be manipulated experimentally to establish causal relationships among the various neuronal changes. The effects of such factors as dietary deficiencies, temperature, radiation, toxic substances, antioxidants, and transmitter-affecting drugs — all known to promote or impede aging processes in other systems[14,114,144,154]

— could be evaluated quite precisely for a variety of specific neurons and neuronal networks in the leech.

In order to suggest a more specific experimental plan, we will review briefly the morphology, physiology, and biochemistry of neurons in the adult leech, and indicate how leech behavior is being studied in terms of these properties. We will outline the development of the leech in some detail, since knowledge of the processes and mechanisms operating during these developmental stages may be essential for understanding senescence, the ultimate phase of life. Indeed, it is possible that some of the changes leading to the physical decline of an organism are extensions of processes that lead first from the egg to the adult. Finally, we will discuss what is known about aging in leeches and what, in our view, remains to be done.

## II. DESCRIPTION OF THE ORGANISM

### A. Behavior and Morphology

Leeches are annelids, i.e., segmented worms. Unlike other annelids, however, they cannot regenerate body parts. Leech body shape varies from long and slender to short and squat (Figure 1). All, however, share many characteristics: (1) they have exactly 32 segments, (2) they are hermaphroditic, devoting one segment to male organs and another to female organs, (3) they have a large sucker on the posterior end, used for attachment and a characteristic "inch-worm" walking, and a smaller sucker on the anterior end, which doubles as a mouth and the second attachment for the walking behavior, and (4) they have external erectile ridges called annuli. There are three orders of leeches, distinguished by whether their mouth is equipped with jaws (gnathobdellids), a proboscis (rhynchobdellids), or a pharynx (pharyngobdellids).[110] All leeches ingest food by suction; most of the gnathobdellids and rhynchobdellids take in only the blood of their prey (i.e., they are "sanguivorous"), whereas the pharyngobdellids also swallow pieces of tissue (i.e., they are "macrophagous").[24] In general, sanguivorous leeches have large gut pouches for storing blood and, consequently, eat very infrequently. The largest species of leech, *Haementeria ghilianii*, eats only every 40 to 100 days, increases its weight up to 10-fold with each feeding, and reaches sexual maturity after only four feedings (about 8 months).[111] Macrophagous leeches consume proportionately less and eat more frequently. For instance, the snail leech *Helobdella triserialis*, will eat 2 to 3 times per week in the laboratory, reaching sexual maturity in about 9 weeks. Adults of the smallest leech species are less than 1 cm in length fully extended, and the largest are nearly half a meter.

The most commonly used leech for physiological studies is the sanguivorous gnathobdellid *Hirudo medicinalis*, the European medicinal leech, mainly for historical and practical reasons (e.g., availability and ease of maintenance, dissection, and recording). Other gnathobdellids, such as *Macrobdella*[54,68] and *Haemopis*[68,151] have very similar nervous systems to that of *Hirudo*. Most developmental studies have been performed on glossophoniids, a family of rhynchobdellids, because their eggs and embryonic cells, being filled with yolk, are quite large, and also because their embryos undergo direct development without a larval stage. The most commonly used glossophonids are *Haementeria ghilianii*[111] and *Helolodella triserialis*.[136] Although the gross structure of the nervous system is somewhat different, many of the neurons in these animals are also homologous to gnathobdellid neurons.[46]

The leech central nervous system (Figure 2A) contains about $10^4$ neurons grouped into 33 neuronal clusters called ganglia. Similar embryonic development occurs in 32 of the ganglia, whereas the most anterior, the supraesophgeal ganglion, has a unique origin. The first 4 of the homologous ganglia form the subesophageal ganglion of the anterior brain, and the last 6 ganglia are fused to form the posterior brain. The other 21 ganglia are nearly identical, and are located ventrally in separate midbody segments. Each midbody ganglion

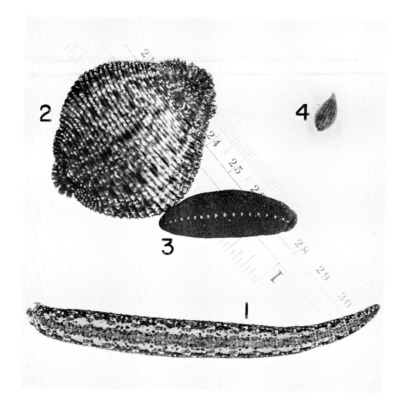

FIGURE 1.    Photograph of four species of leeches used for neurobiological studies. They are (1) *Hirudo medicinalis* (from supplier in France, 4.7 g); (2) *Haementeria ghilianii* (fourth-fed, laboratory-reared, ancestors collected in French Guyana; 115.3 g); (3) *Macrobdella decora* (from supplier in Minnesota; 2.7 g); (4) *Helobdella triserialis* (laboratory-reared; ancestors collected in California; 0.03 g). All animals are sexually mature, but not maximal size.

consists of some 400 neuron somata,[73] most of which are paired bilaterally. In general, all neurons have homologs in each midbody ganglion[88,94,122,123] and some are found even in the brain.[145,148] There are some unpaired neurons[30,134] and some neurons that are present only in a limited number of segments,[124,131] but, to a first approximation, the leech midbody nervous system consists of 21 paired clusters of 200 neurons. Drawings, both dorsal and ventral views, of a typical ganglion in *Hirudo* are shown in Figures 2B and C, showing the locations of the neurons that will be cited in this article.

The neurons in the segmental ganglia connect to one another via three connectives and to the periphery by way of laterally directed segmental nerves. The somata of individual neurons are located in a rind around the outside of the ganglion, where they can be impaled with microelectrodes for electrical recording or for injection with a dye. The structure of a typical motor neuron, injected with horseradish peroxidase and reacted to form a black precipitate, is shown in Figure 2D. Every neuron is monopolar, sending a single branch into the central neuropile to make synaptic contact with other neurons; most neurons also send other branches into one or more connectives or nerves.

## B. Development

### 1. Developmental Stages

Glossiphoniid leeches are particularly well-suited for developmental studies because their eggs and early embryos are large and can be observed, manipulated, and cultured to maturity

99

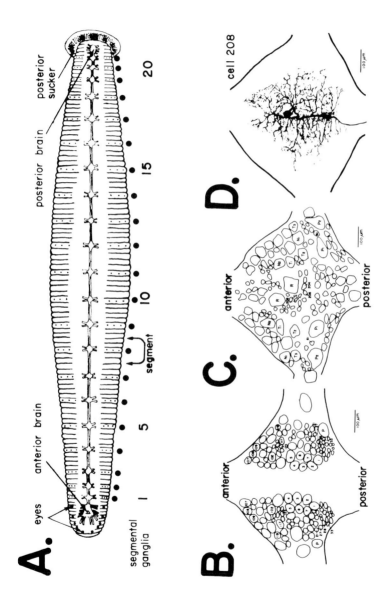

FIGURE 2. Anatomy of the central nervous system (CNS) of the medicinal leech. (A) Diagram of the whole CNS super-imposed on a sketch of the body, viewed from the dorsum. Dots indicate the middle annulus of each midbody segment; the numbers refer to the segmental numbering scheme used throughout this article.[86] (B) The dorsal and (C) the ventral views of a typical midbody ganglion, with cell body locations indicated. The numbers were arbitrarily assigned to facilitate referring to the cells when their functions were determined.[86,94] (D) A camera lucida drawing of a neuron (cell 208, a swimming central pattern generating interneuron), which had been filled with horseradish peroxidase, which was later used to catalyze a reaction that fills the cell with a dark precipitate.[86]

in simple media. In the larger glossiphoniid species, such as *Haementeria ghilianii*, the cells of even the embryonic nervous system are accessible for electrophysiological recording and dye-filling to study developing neurons physiologically and morphologically.[47,61] Smaller species, notably *Helobdella triserialis*, offer the advantages of a shorter generation time (9 weeks), the capacity for a self-fertilizing mode of reproduction, and a smaller size, facilitating light, and electron microscopy. Fortunately, the glossiphoniid leeches are so similar in adult body plan and embryonic development that they are largely interchangeable.

"*Omnia embryogenesis sanguisugae in tres partes divisa est*". All of leech development is divided into three parts: *early embryogenesis* (Figure 3, stages 1 to 8), which extends from cleavage of the fertilized egg to generation of the germinal plate, a sheet of undifferentiated cells oriented longitudinally along the ventral midline of the embryo; *late embryogenesis* (Figure 3, stages 9 to 11), in which the germinal plate cells divide and differentiate into segmental tissues, forming the sexually immature juvenile leech; and *postembryonic development*, which begins with the first feeding and includes the successive maturation of the male and female reproductive systems.[119]

The division of the first two phases of leech embryogenesis into 11 stages is based on studies of *Helobdella triserialis, Theromyzon rude,* and *Haementeria ghilianii.*[28,112,139,140,141] Early embryogenesis begins when large (0.5 mm in *Helobdella*, 2.0 mm in *Haementeria*), yolky eggs are fertilized internally and laid in clutches of transparent cocoons attached to the ventral body wall with up to 200 eggs per clutch in *Helobdella* and over 300 in *Haementeria*. Cleavage does not begin until the egg is laid, thus embryos of a given clutch develop roughly synchronously. The cleavage divisions are invariant and asymmetric (Figure 4), giving rise by stage 6c to an embryo with three endodermal precursors, the A, B and C *macromeres;* one bilateral pair of mesodermal precursors, the M teloblasts; four bilateral pairs of ectodermal precursors, the N, O, P, and Q teloblasts, and a set of *micromeres* of diverse origins.

Each teloblast generates a *germinal bandlet* of several dozen *primary blast cells* by a series of highly asymmetric divisions in which the distal surface of each new primary blast cell remains in contact with the proximal surface of the blast cell formed in the previous division. The germinal bandlets on either side of the embryo merge in stage 7 to form a pair of cell ridges, the left and right *germinal bands.* In each band the ectodermal bandlets (designated by the appropriate lower case letter) lie superficially in the order n, o, p, q, from lateral to medial; the m bandlet lies beneath them. As more blast cells are produced, the left and right germinal bands grow over the dorsal surface of the embryo along crescent-shaped paths and converge at the beginning of stage 8, where the head will form. At the same time, the middle regions of the still-lengthening germinal bands move circumferentially into the ventral surface of the embryo. Eventually during stage 8, the left and right bands coalesce like a zipper at the ventral midline, from future head to tail. The process of coalescence gives rise to the germinal plate, thus ending early embryogenesis. The circumferential migration of the germinal bands reverses the mediolateral order of the ectodermal bandlets so that n is most medial and q most lateral within the germinal plate.

In late embryogenesis (stages 9 to 11) the germinal plate cells proliferate to form the segmental tissues of the adult leech. In the course of this cell proliferation, during stages 9 and 10, the germinal plate thickens and its edges expand circumferentially back into dorsal territory. Eventually the edges meet along the dorsal midline, closing the leech body at the end of stage 10. During stage 9, the germinal plate ectoderm is partitioned into a series of tissue blocks, each corresponding to a future body segment. Segmentation starts at the front of the embryo and progresses rearward until all 32 segments have formed. The formation and segmentation of the nervous system follows the same front-to-rear developmental plan. Primordial segmental ganglia appear first as paired masses of cells on either side of the midline; these masses coalesce to form ganglia containing the approximate adult number of

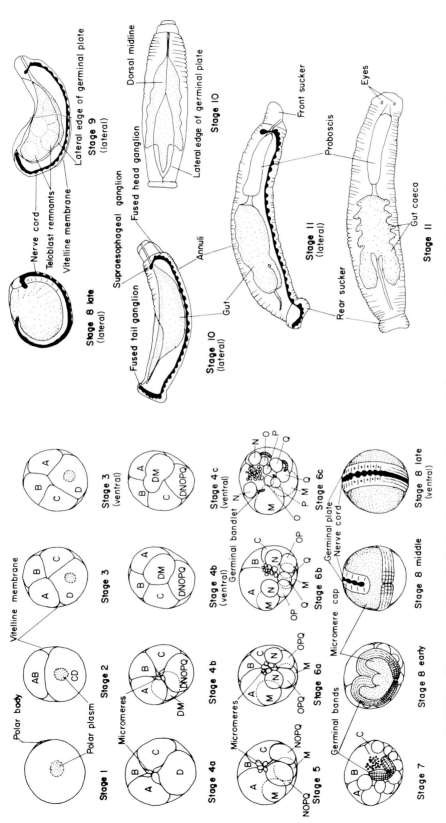

FIGURE 3. Eleven stages of glossiphoiid leech development, beginning with the uncleaved egg. All drawings depict the future dorsal aspect of the embryo unless labeled otherwise. The polar plasm is a region of yolk-free cytoplasm which appears at the dorsal and ventral poles of the egg prior to first cleavage. It is asymmetrically distributed during the early stages of development, going largely to the D macromere in stage 4a, and thence to the five pairs of teloblasts.

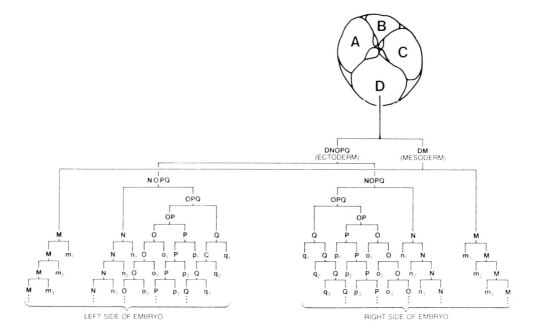

FIGURE 4.    The pedigree of the teloblasts and primary blast cells, starting with the D macromere in stage 4a. Note that, whereas the N and Q teloblasts are fundamentally distinguishable because their sister blastomeres, OPQ and OP, go on to cleave further, the O and P teloblasts arise by a symmetric, terminal cleavage and are indistinguishable except by position.

cells. These cells grow neurites that form the ganglionic neuropil, the interganglionic connectives, and the segmental nerves to the body wall and viscera. As development continues, the 21 midbody ganglia move apart, leaving the 7 hindmost and 4 frontmost ganglia fused as the tail brain and subesophageal ganglion of the head brain, respectively. Meanwhile, the gut has formed around the yolky remnants of the macromeres and teloblasts, which provide nutrition during embryogenesis. By the end of stage 11, also the end of late embryogenesis, these remnants have been consumed and the juvenile leech is ready for its first meal, thus beginning postembryonic development.

### 2. Cell Lineages

Further details as to the development of the nervous system from the teloblasts have been obtained using horseradish peroxidase (HRP) or fluorescent dyes covalently linked to synthetic peptides as cell lineage tracers which can be injected into identified blastomeres in the leech embryo.[138,139] These substances appear not to affect subsequent development and remain confined within the injected cell and all its descendants. These techniques have shown that the divisions of the primary blast cells begin even before the germinal bands coalesce into the germinal plate. This is especially obvious within the m bandlet where clusters of cells, primordial mesodermal half-somites derived from single primary blast cells, can be seen in the anterior portions of the germinal bands in stage 7.[150]

The cell lineage tracers persist as late as the end of stage 10 in *Helobdella* at which time progeny of the labeled blastomere can be identified by histochemical staining or fluorescence microscopy. Thus it has been shown that each teloblast gives rise to a distribution of progeny sufficiently idiosyncratic and invariant from segment to segment so that a distinct and recognizable pattern in the stage 10 embryo can be associated with each teloblast (Figure 4). In general, the progeny of each teloblast are found only ipsilateral to it in the germinal plate, and when a teloblast is injected with tracer after the initiation of blast cell production, there is a sharp boundary between unlabeled (anterior) tissue and labeled (posterior) tissue. Thus, in normal development there is little or no migration of cell bodies from side to side across the ganglionic midline or from posterior to anterior between adjacent ganglia, so that the crossed and uncrossed projections of various neurons must arise from axonal growth, not from cell body migration.

Reconstruction of the total neuronal staining pattern within the ganglion from serial sections of embryos, such as those shown in Figure 5, reveal a unique topography of the major ganglionic subpopulation arising from each teloblast (Figure 6). These findings suggest that the developmental cell lineages represented by the descendants of each teloblast correspond to distinct, identifiable, neuronal kinship groups. In view of the positional invariance of identified neurons in the segmental ganglia, it would appear that each identified neuron is the lineal descendant of a particular teloblast. By similar techniques, it has been shown that the supraesophageal ganglion of the head brain and the rest of the most anterior body wall (together constituting the *prostomial ectoderm*) derive not from the ectoblasts at all, but rather from the micromeres produced by the asymmetric third cleavage in stage 4a.[137,139]

These and many other unpublished observations strongly suggest that there is considerable determinacy in leech neurodevelopment, and in particular that identifiable cells in adult ganglia normally arise from particular teloblasts. However, ablation experiments, in which particular teloblasts are killed by injection of pronase or DNase, coupled with dye-filling of teloblasts, show that some cells in the developing nervous system have a limited ability to change their position on their lineage in response to loss of other cells.[6,135]

### 3. Neuronal Physiology and Morphology in Development

In *Haementeria*, the germinal plate and ventral nerve cord can be dissected from the embryo as early as stage 8, and individual cells impaled with microelectrodes even before they develop neurites. This has permitted the description of the electrophysiology and morphology of identified neurons, especially the pressure-sensitive sensory neurons, from their birth through functional maturity.[47,61] Thus far, it has been shown that the peripheral processes of these cells develop after the central ones, that the cells are electrically excitable before their sensory endings are completely established, and that synaptic potentials are first seen about a day after electrical excitability is established. Furthermore, pressure-sensitive neurons (P cells) that ultimately subserve dorsal sensory fields make only sparse, inappropriate arborizations in ventral fields, even though they grow across future ventral areas that are not yet fully colonized by the ventral P cell. This difference in initial branching may result from inherent differences between the two cells or from differences in the substrates they traverse as they leave the ganglion. Also, different processes of the same sensory neuron recognize and avoid overlapping one another, even though they do overlap processes of other pressure-sensitive cells. This self-avoidance seems intrinsic to the neuron and was not impressed upon it by cues in the substrate, as was shown by experiments in which individual nerve roots were damaged in embryonic *Haementeria*.[45a] After a few days of subsequent development, sensory cell arborizations were examined by dye filling. It was found that the arborizations of processes normally present in the damaged root were missing or greatly reduced, while other processes from the same cell, reaching the periphery by other routes, had greatly expanded arborizations which filled the vacant space, still without self-overlap.

### 4. Biochemical Development

Radiochemical techniques were used to examine the capacity of *Haementeria* embryos to synthesize and accumulate acetylcholine, serotonin, and GABA from labeled precursors.[15,129] Acetylcholine synthesis is very low when the first axons grow out during stage 8, but increases rapidly, being 25-fold higher by the end of stage 10. Other transmitters seem to follow the same time course. Serotonin and dopamine can be demonstrated in individual

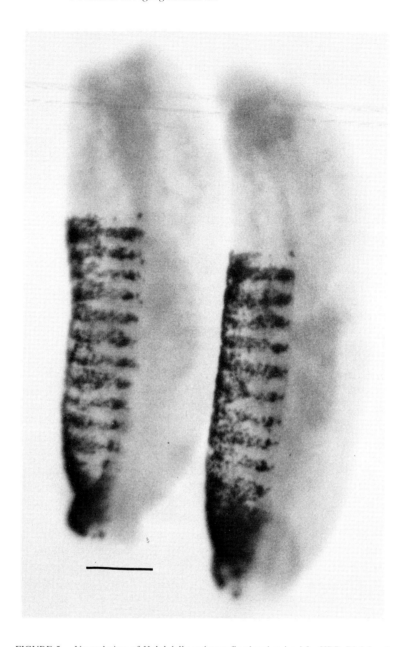

FIGURE 5.    Ventral view of *Helobdella* embryos fixed and stained for HRP. Righthand P teloblasts were injected at stage 7, after some primary blast cells had already been produced. Note, therefore, that the anterior part of the embryo, derived partly from those earlier blast cells, is unstained. In the posterior part, the labeled progeny of the P teloblast are confined to the right half of the embryo (although, in this view, it's on the left) and form a pattern that is repeated from segment to segment and from animal to animal. Anterior is up; scale bar, about 100 μm.

neurons in the leech nervous system using glyoxylic acid-induced fluorescence. Moreover, the serotonergic neurons can be ablated selectively by bathing intact embryos in the toxic serotonin analog 5,7-dihydroxytryptamine.[36] Treated embryos develop normally except that they fail to exhibit swimming behavior. These leeches have not lost the ability to swim because bath-applied serotonin restores swimming. Putative GABA neurons in adult and

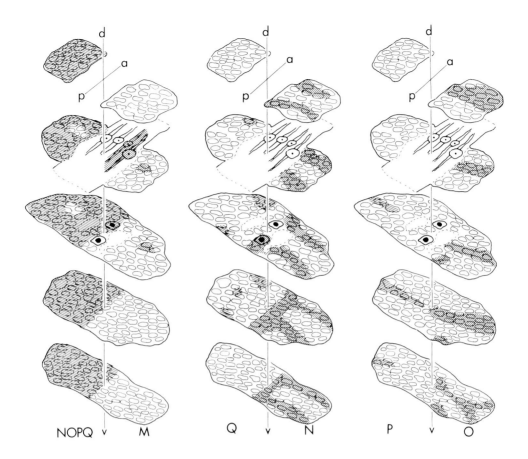

FIGURE 6. Embryonic origins of the cells of the stage 10 *Helobdella* ganglion. The drawings show five horizontal sections through a midbody segmental ganglion, dorsal (d) aspect at the top, anterior (a) edge facing away from the viewer. The two pairs of dark, elongated contours in the center of the second section from the top represent identifiable muscle cells in the longitudinal nerve tract. The two dark, circular contours in the center of the middle section represent the neuropil glia. The other, faint contours do not correspond to actual cells but are shown to indicate the approximate size, disposition, and number of neurons in the ganglion. In each half-ganglion, domains shown crosshatched contain descendants of the teloblasts or blastomere indicated.

embryonic leeches have been identified autoradiographically by their selective uptake of [3]H-GABA.[15] Moreover, the functional identification of those cells which take up GABA is possible by filling electrophysiologically identified cells with lucifer-yellow or HRP prior to incubating them with [3]H-GABA.

Recently, monoclonal antibodies have been produced after immunizing a mouse with leech nerve cord tissue.[153] By reacting the antibodies against adult leech ganglia, it was found that many were specific to a limited number of nerve cells, one being specific to just the N mechanosensory neurons. By determining the onset of sensitivity to these antibodies during neurogenesis, it should be possible to determine when the antigenic molecules first appear. Those molecular moieties that appear during appropriate stages could be involved in axonal outgrowth, pathfinding, or synapse formation. These antibodies would have great potential for finding the molecular basis of neurogenesis.

## C. Physiological Properties of Neurons

On the basis of morphological and physiological properties, about 25% of the neurons in each midbody ganglion have been identified as individuals.[86] Such features as the resting potential, the size and shape of the action potential, and the ability to produce prolonged bursts of impulses can distinguish among some neurons.[88] However, the branching pattern of a neuron, its connections to other cells, and its activity during behavior must be taken into account to positively identify many neurons.[33,94,123,124,134]

There are seven mechanosensory neurons on each side of each segmental ganglion:[88] three T (touch) cells, two P (pressure) cells, and two N (nociceptive, i.e., pain) cells (see Figure 2C). The T cells divide half the segmental body wall into three receptive fields: dorsal, lateral, and ventral. The P cells divide it into a doral and ventral field, and the N cells have very complex fields.[5] The other known sensory receptors for light[48] and water wave detection,[31] are located in sensilla, small sensory patches in characteristic locations in the skin.

Overt body movements are controlled by five sets of muscles in the body wall and skin.[122] Two of the sets are the dorsoventral muscles (whose contraction causes flattening of the body) and the longitudinal muscles (whose contraction causes shortening of the body). With just these muscles, a leech is able to perform a variety of behavioral acts. Local bending, shortening, and swimming will be discussed subsequently. This differential control is effected by activating the motor neurons in different combinations and in different temporal sequences (Figure 7). Most, if not all, of the motor neurons to these muscles have been identified.[94,122] In addition, the motor neurons causing the contractions of the heart tubes have been identified.[123]

There are also two somewhat unusual types of effector neurons. The first are the peripheral inhibitors, which are essentially motor neurons whose transmitter causes hyperpolarization of the muscle membrane potential.[122] These peripheral inhibitors also make synaptic contacts within the central nervous system, inhibiting the excitatory motor neurons to the same muscles contacted by the inhibitors.[94] The second unusual type of effector are the Retzius cells, the largest two neurons in any segmental ganglion. Activity of these cells causes release of mucus from the skin,[67] increased relaxation rate in muscles,[79] and a propensity for the nervous system to produce the swimming activity pattern.[143]

Chemical synaptic potentials, both excitatory and inhibitory, have been characterized among identified neurons.[83] Some synaptic potentials facilitate and others depress with repeated use. In fact, the same mechanosensory cell makes facilitating contacts with one motor neuron and depressing contacts with another.[85] The quantal basis for transmitter release[91] and the mechanism of facilitation of transmitter release[90,125] have been established for the synapse between the interneurons generating the contractile rhythm of the heart (the HN cells) and the heart tube motor neurons (the HE cells). Many electrical contacts have been described; most constitute coupling between bilateral homologs,[22,39,43,70,94,120] but some are from sensory cells to interneurons,[87] others are from sensory cells to motor neurons,[89] and still others are between synergistic motorneurons.[94]

Leech glial cells were the first to be investigated physiologically in any detail. Their membrane potential is unresponsive to electrical or chemical stimulation, responding only to changes in the external potassium concentration.[59] The glial cells present a minimal barrier to diffusion of ions, although larger molecules apparently pass through the glial cells before having access to the neurons.[58] Hence, leech glial cells, like those in other nervous systems, appear not to be directly involved in electrical function.

**D. Neurotransmitters**

A variety of chemical, pharmacological, and histochemical techniques have indicated the presence of a number of transmitters in the leech segmental ganglion and, in many cases, the particular neurons containing the transmitter have been identified. Using the ''hot zap'' technique whereby ganglia were incubated in a mixture of isotopically labeled transmitter precursors, then analyzed for labeled transmitters, leech neurons were found capable of synthesizing acetylcholine (ACh), serotonin, gamma-amino butyric acid (GABA), dopamine, and octopamine.[106]

ACh appears to be the transmitter used by the excitatory motor neurons to the body wall muscles; it is synthesized in these neurons.[107] Nicotinic cholinergic agonists and antagonists

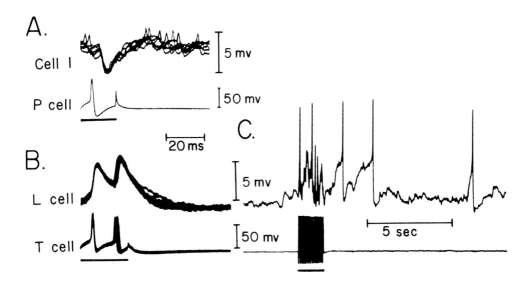

FIGURE 7.   Electrophysiological properties of leech neurons. (A) Intracellular recordings from a mechanosensory neuron (P cell) and an inhibitor of the dorsal longitudinal muscles (cell 1). These are actually five superimposed recordings, which begin when depolarizing current was passed into the P cell to cause a single impulse which, about 5 msec later, produces an inhibitory chemical synaptic potential in cell 1. The short and constant latency strongly suggests, but does not prove, that this connection is monosynaptic. (B) Similar recordings from another mechanosensory neuron (T cell) and the excitatory motoneuron to both dorsal and ventral longitudinal muscles (L cell), the motorneuron primarily responsible for the shortening response. Each of the five superimposed traces began at the start of a stimulus that produces two impulses in the T cell which, in turn, produce excitatory synaptic potentials occurring with almost no latency. This short latency, and other criteria,[89] establish this as an electrical synaptic potential. The small upward deflections in the cell 1 recording are impulses, they are small because they are generated at a distance from the site of recording in the cell body. (C) Recordings from the same two neurons as in (B) with the same voltage scale but a compressed time scale. The T cell was stimulated intracellularly to produce 15 impulses in 1 second, mimicking the response to a moderately light touch to the skin. Note the barrage of synaptic potentials elicited in the L cell, some of which cause impulses. Some of the synaptic potentials, particularly those 1 to 7 sec after the last T cell impulse, must be generated by other neurons, probably interneurons, that are also activated by the T cell stimulation. Below the T cell and P cell recordings, bars are drawn to indicate the time when depolarizing current was passed into the cell.

are effective at the body wall neuromuscular junctions[57] and the motor neurons have a 10-fold greater concentration of acetylcholine esterase (AChE) in their cytoplasm than do any other neurons.[129]

Serotonin is strongly implicated as the transmitter in the Retzius cells and five to seven other identified neurons. These neurons have the characteristic yellow histofluorescence using either the formaldehyde[104] or glyoxylic acid[69] fixation technique. Chemical analysis of individual neurons show a high concentration of serotonin in these cells.[71] The effects of stimulating these neurons can be mimicked by application of serotonin.[67,79,93,143] These neurons are selectively killed by the toxic serotonin analog 5,7-dihydroxytryptamine[36] and serotonin is released in a Ca-dependent manner by stimulation of the Retzius cells.[143]

The case for the other putative transmitters is far less developed. Two or three pairs of neurons (depending upon the species) show the typical histofluorescence for dopamine.[104] A number of neurons have a high affinity uptake system for GABA[15] and a small number of neurons (in the segments containing the male and female sexual organs and in the supraesophageal brain ganglion) react to the antibody for leu-enkephalin.[152] Tests for other peptides or any amino acids have not yet been performed.[3,129]

## E. Regeneration of Synaptic Connections

The synaptic input from specific mechanosensory neurons in one ganglion onto their target motor neurons in an adjacent ganglion can be removed by cutting or crushing one of the connectives between the ganglia. After such lesions, the sensory neurons regrow their severed axons and make contact with the same motor neurons they had contacted previously.[41] This specificity of reconnection remains precise when the ganglia are maintained in culture for the 3 to 6 weeks required for the regeneration to occur.[82] The ability to regenerate processes and even some degree of specificity, is retained even after pairs of neurons are removed from the ganglion and grown together in tissue culture.[99]

The S cell, an interganglionic interneuron, will sprout a number of processes at its severed end after a connective is cut. These processes make contact with their own distal stump, which can survive for many weeks, and use it as a scaffolding to re-establish contact with its normal target, the process of the S cell from the adjacent ganglion.[13,84] This reconnection still takes place readily after the giant glial cells in the connective have been destroyed.[25] Mechanosensory and motor neurons re-establish their peripheral innervation patterns after a nerve is cut or crushed, possibly by a mechanism similar to that used by the regenerating S cell processes.[128]

After all the connectives to a segmental ganglion are cut, that body segment gains the ability to produce swimming movements after about a week.[53] This behavioral change probably results from a strengthening of synaptic pathways,[49] possibly by sprouting of the processes of various swim-related neurons.[37] This suggests that sprouting can be triggered in processes other than the one that is cut, and that sprouting may help to compensate for functional loss after damage to the nervous system.

## F. Neuronal Control of Behavior

A variety of leech behaviors have been described.[35,38,78,109,111] At least the beginnings of a complete neuronal explanation is available for some of these behavioral acts: shortening,[74,89] swimming,[116,117] heartbeat control,[12,116,118] and local bending.[51] A short description of three of these analyses will be given to show the current status of this type of research.

### 1. Shortening

If a leech is touched or pinched, especially near its front end, it often shortens its whole body[54] by contracting all its longitudinal muscles nearly simultaneously.[122] As indicated in Figure 8A, this contraction is caused principally by the activation of the paired L motor neurons in every segment, both by direct synaptic contact from the mechanosensory neurons[89] and by way of intersegmental interneurons, such as the unpaired S cell.[45,74] The S cell in each ganglion makes electrical contact with the S cells in each adjacent ganglia,[30] making a fast through-conducting system that runs the length of the animal. The S cell also makes electrical contact with the L cell in its own ganglion[34] and is activated by mechanical stimuli and changes in light levels.[2,65] The pathway from the mechanosensory T cells to the S cell, which appears to be a monosynaptic electrical connection, actually has a "coupling inter-neuron" interposed between them.[87] Although there are other, as yet unidentified, inter-neurons involved in this behavior, the identified four neuron reflex arc constitutes a significant part of the shortening network.

### 2. Swimming

In response to mechanical stimulation,[54,149] many leech species swim. They stretch out to full length and flatten by contracting their dorsoventral muscles, then undulate up and down by contracting their dorsal and ventral longitudinal muscles alternately.[55] Forward thrust is generated by the contraction of each set of muscles at a slightly later time in each more posterior segment, this produces a front-to-back undulatory wave that pushes back on

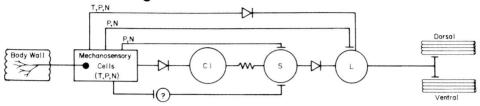

# A. Shortening

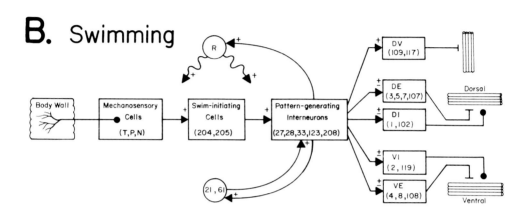

# B. Swimming

# C. Heartbeat

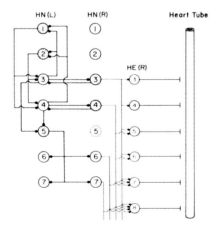

FIGURE 8. Neuronal circuits producing three different leech behaviors: (A) shortening, (B) swimming, and (C) heartbeat. In all cases, a circle represents a single cell and a square represents a class of cells. Lines eminating from these squares and boxes represent neuronal branches. These terminate in symbols representing: monosynaptic chemical excitation (a T junction) or inhibition (a filled circle); electrical contact, either bidirectional (the resistor symbol) or one-way (the diode, or rectifier, symbol); or polysynaptic (arrowheads), either excitatory (+) or inhibitory (−), or both. (A) CI = coupling interneuron; S = S cell; L = L cell. ? = unidentified interneuron whose presence is known from recordings in the S cell. On the right are indicated dorsal and ventral longitudinal muscles. (B) DV = dorsoventral, or flattener, motorneurons; R = Retizus cell; numbers = identified neurons of that class (see Figure 2). The same dorsal and ventral longitudinal muscles are indicated, as well as the flattener muscle. (C) HN = heart interneuron; HE = heart excitor; L = left (i.e., located on the left side of the ganglion); R = right; numbers indicate the midbody ganglion in which that cell is located. The shaded cells are those which are in the same state (i.e., firing or not firing impulses) at the same time; the unshaded cells are in the opposite state from the shaded cells.

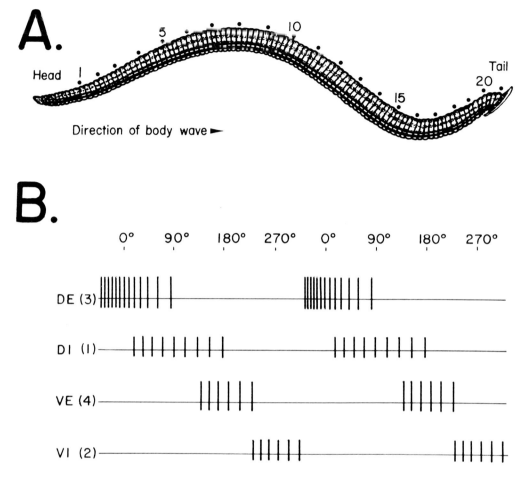

FIGURE 9. Leech swimming. (A) Drawing of a leech, swimming from right to left, with the segments numbered as in Figure 2. (B) Idealized impulse pattern from four motorneurons during swimming. DE = dorsal longitudinal muscle excitatory motorneuron; DI = dorsal inhibitor; VE = ventral excitor; VI = ventral inhibitory. The parenthesized numbers are the particular motorneurons of each class whose firing patterns are sketched. Each cycle is arbitrarily taken to begin with the middle DE spike, which thus determines 0° phase. The next DE burst finishes a cycle (i.e., 360°) and begins a new one (i.e., 0°). Each cycle is divided into 4 parts, which roughly correspond to the midpoints of the impulse bursts of each of the motorneurons. The actual cycle period varies from 400 to 2000 ms.[55]

the water and propels the leech forward (Figure 9A). A swimming episode consists of 5 to 50 repeated cycles of this undulation; each cycle lasts 500 to 2000 msec. These movements are caused by a repeated four-phase cycle of impulse bursts in the dorsal and ventral motor neurons in each segment (Figure 9B).[56] This same pattern can be elicited by the isolated ventral nerve cord,[52] thereby proving that the swimming rhythm, like many other rhythmic behaviors in both vertebrates and invertebrates[20,50] is generated by a central pattern generator (CPG) without the need for sensory feedback.

Although this behavior is far from being completely characterized in neuronal terms, the contributions of many of the neurons at different hierarchical levels have been characterized[116] (Figure 8B). There is a group of pattern-generating neurons whose connections (1) to each other within each ganglion produce the elemental swimming rhythm,[32,33] (2) to each other between segments produce the undulatory wave,[33,130] and (3) to the appropriate motor neurons produce the impulse bursts that contract the muscles properly.[98,133] This pattern-generating network is activated by unpaired swim-initiating neurons (cells 204 and 205) in all but the first eight ganglia.[131,132] Intracellularly activating any swim-initiating neuron excites all the others up and down the nerve cord, thereby activating the pattern-generating cells. There is an excitatory feedback loop between the swim-initiators and the pattern-generators that helps to prolong the swimming episode. The swim initiators are also strongly excited by mechanosensory neurons which initiate swimming.

The tendency for leeches to swim is controlled by the level of serotonin in the blood bathing in the nervous system.[143] The site of action of the blood-borne serotonin is not known, but the primary source of it appears to be the Retzius cells. Other serotonin-containing neurons appear to have a much more direct, seemingly synaptic effect on the swim-generating system.[93] This dual synaptic and hormonal effect of the same substance is similar to the dual effects of peptide hormones that appear to be transmitters in neuronal pathways synergistic with their hormonal functions.[5] In the leech, the cellular mechanisms of these differences in effects of serotonin are being studied directly.

### 3. Heartbeat

Unlike the previous two behaviors discussed, heartbeat is an ongoing rather than a transient motor act. Gnathobdellid leeches have two contractile heart tubes, one on each side, which beat in a characteristic pattern. One tube contracts synchronously over its whole length, whereas the other produces a back-to-front peristaltic wave. This pattern repeats at 15 to 30 sec intervals, with the left and right sides taking turns as the peristaltic and synchronous sides every 10 to 50 cycles.[12,123] These contractions are caused by periodic bursts of impulses in heart motor neurons, the HE cells, a pair of which are located in each ganglion. Left to their own devices, these neurons would produce impulses at a constant rate, instead, they produce impulse bursts because they receive periodic barrages of very strong inhibitory potentials from interneurons, the HN cells.[124] Like the HE cells, there are only two HN cells per ganglion; however, the HN cells are present in only the first seven ganglia. The HN cell activity pattern results from their own inherent rhythmicity, coupled with connections among them (Figure 8C).

To gain an appreciation of the generation of the heartbeat pattern, consider the HN(3) and HN(4) pairs on each side as being pivotal. Their reciprocally inhibitory connections with HN(1) and HN(2) on the same side sets the basic rhythm; their reciprocally inhibitory connections with HN(3) and HN(4) on the other side assure that the two sides are never active at the same time; and the inhibition of HN(5) determines whether that side is the synchronous or the peristaltic side.[12,118] The key to whether the pattern is synchronous or peristaltic on a given side is the activity of the HN(5) cells. During any given heartbeat cycle, one of them is totally silent and the other fires during the time that HN(3) and HN(4) on the same side are silent. The side with the silent HN(5) is the synchronous side, whereas the side with the periodically active HN(5) is the peristaltic side. The only element lacking for a complete description of the generation of the heartbeat pattern is the cell or cells that determine which HN(5) is silent.

Although the systems generating these three behaviors are quite different, some generalities are possible. Swimming and shortening are both episodic behaviors, being triggered by mechanosensory stimuli delivered to different body locations. Swimming and heartbeat are both rhythmic, but they have very different cycle periods (about 1 sec for swimming and 20 sec for heartbeat). The swimming pattern is almost certainly generated exclusively by connections among interneurons, whereas heartbeat results from both inherent oscillatory properties of the interneurons and the connections among them. At least one new neuronal property — presynaptic facilitation by depolarization of the presynaptic terminals — was discovered because a particular feature of the motor pattern could not be explained by conventional synaptic mechanisms.[125]

## III. AGING IN LEECHES

### A. What is Known

In contrast to the wealth of knowledge about the leech nervous system through development to maturity, there is almost no information about changes after adulthood as the animal ages. However, there are changes in the behavior of leeches in the course of their lives,[111] which implies that some neuronal properties must change with age. Given its advantages as an experimental preparation, the leech nervous system would appear to be an excellent place to seek such changes.

To know when senescence begins, one must first know the normal course of an animal's life. The normal life span of various leech species, both in the wild and in captivity, ranges from less than 1 to 3 or 4 years.[24] Some species have highly variable life spans, ranging from 1 to 3 years depending upon such factors as whether they reproduce, local temperature, and rainfall.[108,126] In general, the normal lifespan for many species is thought to be 2 years, although survivorship curves exist only for one species, *Glossiphonia complanata*.[76] A variety of changes in apparently older leeches have been noted, as discussed below.

*1. Morphological Changes*

In many invertebrates, size and appearance of external features show changes with age.[14,96] In leeches there are few, if any external signs of organ senescence and degeneration,[127] but size and development of reproductive organs have been used to identify age cohorts in many studies.[66,75,147] Fibrous degeneration has been observed in the ovaries in one species prior to death.[127] In contrast, spermatogenesis and testicular integrity is maintained just prior to death in another species.[126] As in other invertebrates,[44,80] lipofuscin accumulation occurs in leeches.[27] Also, juvenile *Hirudo medicinalis* accumulate an iron-containing hemoglobin breakdown product in their botryoidal tissue (a diffuse storage and digestive tissue), whereas the adjacent vas-fibrous tissue acquires pigment granules related to bile pigment.[10] In other species, the adipose cells surrounding the digestive tract fulfill this function, becoming progressively filled with "bile pigment" at the expense of fat.[8] However, it is possible that the accumulation of these pigments is more related to maturation and feeding rather than to senescence.

Many authors have commented that size and growth rates, which may be indices of the age of the animal, are affected by environmental factors such as temperature[19] and availability of food.[1] Steady weight gains with age have been observed in several macrophagous leeches[23,40,75] whereas sanguivorous leeches, with their large blood-storing gut pouches, display a markedly saltatory growth pattern with sharp weight gains following a blood meal.[29,111] However, even under laboratory conditions, 4th-fed *Haementeria ghilianii* may show considerable weight variation due to differential amounts of blood ingested at each feeding.[111] Feeding also affects sexual maturation, reproduction, and pigment accumulation in this species.

*2. Behavioral Changes*

As in other invertebrates,[7,81,96] leeches show a diminished activity level and responsiveness to external stimuli as they age.[111] Young leeches show more pronounced foraging behavior, including exploratory movements and swimming. With age and each successive feeding the foraging behavior is progressively diminished, culminating in the complete absence of swimming and body waving in 3rd- and 4th-fed adults which have just reached sexual maturity. It would be interesting to ascertain whether such age-related behavioral changes are a result of inhibition or functional debility at the neuronal level. It is also necessary to extend these observations to older animals to describe the behavioral changes that accompany senescence.

### 3. Reproduction

There is an unfortunate lack of information regarding the functional decline of leech reproductive processes with age, although such declines are commonly found in a variety of other animals.[100] Offspring development and viability as well as the number of eggs deposited may be influenced by maternal age. The observation that egg production by glossiphonids declines in captivity[142] indicates environmental involvement in these processes.

Although both *Hirudo medicinalis* and *Haementeria ghilianii* produce multiple broods throughout their life span, many species die after breeding.[1,4,9,76,77,108,109] Some species breed once or twice before death, depending on environmental conditions.[1,19,60,66,75,76,126] In other species, only those individuals that fail to reach sexual maturity with the rest of their cohort survive the winter and breed the following year.[23,40,75] It would be interesting to investigate how environmental cues in such organisms lead either to continued vitality or to death, and the possible mediating role of the nervous system.

### B. What Needs to be Done

Because the life cycles of at least some species of leeches are sensitive to environment, in order to study aging reliably in any species, it is necessary first to investigate its survivorship in a constant environment.

The neuronal networks controlling some leech behavioral patterns (i.e., swimming) show some significant age-related alterations; these might be detected as electrophysiological deficits in neurons. First the changes related to senescence would have to be separated from those related to maturation. Then physiological changes in neurons would be sought as senescence occurs, followed by the investigation of such changes as lipofuscin accumulation or disruption of neurotransmission in physiologically deficient neurons. Finally, an attempt could be made to induce these changes exogenously — by dietary deficiency, temperature, radiation, or antioxidants — to determine whether these changes cause senescence. Other senescent changes may occur in sensory systems such as the T, P, and N cells. Both the development and the electrophysiology of sensory and motor circuits has been studied previously. Further work may establish whether functioning of these circuits does decline with age and whether these changes contribute to the decline in the behavior of the animal.

The particular strength of using the leech nervous system in aging studies is that the changes in a neuronal network can be studied electrophysiologically, cytologically, and biochemically at the level of the individual cell. In this way, it should be possible to ascribe particular age-related functional deficits to a chain of morphological and biochemical events that occur in particular cells, and then characterize the events that take place in these cells. In the leech, the relation of the aging process to changes in individual neurons would be much more apparent than in more complex nervous systems, in which redundancy of nervous system function can compensate for large deficits in neuronal function. In this way, the inherent strengths of the leech as an experimental preparation and the wealth of information available about the function of its nervous system could be brought to bear on the causes of the deterioration of nervous system function in aging.

## ACKNOWLEDGMENT

We would like to thank Gunther Stent for intellectual inspiration over the years, and for reading an earlier version of this manuscript. We would also like to thank Ruben Morfin for redrawing several of the figures used herein. Much of the research discussed has been funded by NSF and NIH research grants to William Kristan, NIH and March of Dimes support to David Weisblat, as well as NIH, NSF, and March of Dimes research grants to Gunther Stent.

# REFERENCES

1. **Aston, R, J. and Brown, D. J. A.,** Local and seasonal variations in populations of the leech *Erpobdella octoculata* (L) in a polluted river warmed by condensor effluents, *Hydrobiologia,* 47, 347, 1975.
2. **Bagnoli, P., Brunelli, M., and Magni, F.,** A fast conducting system in the central nervous system of the leech *Hirudo medicinalis, Arch. Ital. Biol.,* 110, 35, 1972.
3. **Barker, J. L. and Smith, T. G., Jr., Eds.,** *The Role of Peptides in Neuron Function,* Marcel Dekker, N.Y., 1980.
4. **Becker, C. D. and Katz, M.,** Distribution ecology and biology of the salmonid leech *Piscicola salmositica* (Rhynchobdellae, Piscicolidae), *J. Fish. Res. Board Can.,* 22, 1175, 1965.
5. **Blackshaw, S.,** Morphology and distribution of touch cell terminals in the skin of the leech, *J. Physiol.,* 320, 219, 1981.
6. **Blair, S. S. and Weisblat, D. A.,** Ectodermal interactions during neurogenesis in the Glossiphoniid leech *Helobdella triserialis, Dev. Biol.,* 91, 64, 1982.
7. **Blest, A. D.,** Longevity, palatability and natural selection in five species of new world saturniid moths, *Nature (London),* 197, 1183, 1963.
8. **Bobin, S.,** Sur les cellules a sphérules colorées et leur parenté avec les cellules adipeuses chez *Glossosiphonia complanata* L. (Hirudinée Rhynchobdelle), *Arch. Zool. Exp. Gen.,* 87, 69, 1950.
9. **Bouvet, J.,** Notes sur les Hirudinées des Alpes Francaises. III. *Erpobdella octoculata* L. (Sous-ordre des Pharyngobdellae, famille des Erpobdellidae), *Trav. Lab. Hydrobiol. Piscic. Univ. Grenoble,* 66/68, 89, 1977.
10. **Bradbury, S.,** The botyroidal and vaso-fibrous tissue of the leech *Hirudo medicinalis, Q. J. Microsc. Sci.,* 100, 483, 1959.
11. **Brody, H. J.,** Organization of the cerebral cortex. II. A study of aging in the cerebral cortex, *J. Comp. Neurol.,* 102, 511, 1955.
12. **Calabrese, R. L.,** Neural generation of the peristaltic and nonperistaltic heartbeat coordination modes of the leech *Hirudo medicinali, Am. Zool.,* 19, 87, 1979.
13. **Carbonetto, S. and Muller, K. J.,** A regenerating neurone in the leech can form an electrical synapse on its severed axon segment, *Nature (London),* 267, 450, 1977.
14. **Clark, A. M. and Rockstein, M.,** Aging in insects, in *The Physiology of Insecta,* Rockstein, M., Ed., Academic Press, N.Y., 1964.
15. **Cline, H. T.,** ³H-GABA uptake selectively labels identifiable neurons in the leech nervous system, *J. Comp. Neurol.,* 215, 351, 1983.
16. **Coleman, P. D. and Goldman, G.,** Neuron count in the locus coeruleus of aging rat, in *Aging,* Vol. 17, Raven Press, N.Y., 1981, 23.
17. **Corso, J. F.,** Sensory processes in man during maturity and senscence, in *Neurobioloy of Aging,* Ordy, J. M. and Buzzie, K. R., Eds., Plenum Press, N.Y., 1975, 119.
18. **Davies, I. and Fotheringham, A. P.,** Lipofuscin — does it affect cellular performance?, *Exp. Gerontol.,* 16, 119, 1981.
19. **Davies, R. W. and Reynoldson, T. B.,** A comparison of the life cycle of *Helobdella stagnalis* (Linnaeus 1758) (Hirudinoidea) in two different geographical areas in Canada, *J. Anim. Ecol.,* 45, 457, 1976.
20. **Delcomyn, F.,** Neural basis of rhythmic behavior in animals, *Science,* 210, 492, 1980.
21. **Drachman, D. A.,** The cholinergic system, memory and aging, in *Brain Neurotransmitters and Receptors in Aging and Age-Related Disorders,* Enna, S. S., Samorajski, T., and Beer, B., Eds., Raven Press, N.Y., 1981, 255.
22. **Eckert, R.,** Electrical interaction of paired ganglion cells in the leech, *J. Gen. Physiol.,* 46, 573, 1963.
23. **Elliott, J. M.,** The life cycle and production of the leech *Erpobdella octoculata* (L). (Hirudinea:Erpobdellidae) in a lake district stream, *J. Anim. Ecol.,* 42, 435, 1973.
24. **Elliott, J. M. and Mann, K. H.,** A key to freshwater leeches, *Freshwater Biol. Assoc. Sci. Publ.,* 40, 1, 1979.
25. **Elliott, E. J. and Muller, K. J.,** Synapses between neurons regenerate after destruction of ensheathing glial cells in the leech, *Science,* 215, 1260, 1982.
26. **Enna, S. S. and Strong, R.,** Age-related alterations in central nervous system neurotransmitter receptor binding, in *Brain Neurotransmitters and Receptors in Aging and Age-Rlated Disorders,* Enna, S. S., Samorajski, T., and Beer, B., Eds., Raven Press, N.Y., 1981, 133.
27. **Fawcett, D. W.,** *The Cell,* W. B. Saunders, Philadelphia, 1966, 285.
28. **Fernandez, J.,** Embryonic development of the Glossiphoniid leech *Theromyzon rude:* characterization of developmental stages, *Dev. Biol.,* 76, 245, 1980.
29. **Fjeldsa, J.,** Records of *Theromyzon maculosum* (Rathke 1862) Hirudinea in N. Norway, *Norw. J. Zool.,* 20, 19, 1972.

30. **Frank, E., Jansen, J. K. S., and Rinvik, E.,** A multisomatic axon in the central nervous system of the leech, *J. Comp. Neurol.,* 159, 1, 1975.
31. **Friesen, W. O.,** Physiology of water motor detection in the medicinal leech, *J. Exp. Biol.,* 92, 255, 1981.
32. **Friesen, W. O. and Stent, G. S.,** Generation of a locomotory rhythm by a neural network with recurrent cyclic inhibition, *Biol. Cybern.,* 28, 27, 1977.
33. **Friesen, W. O., Poon, M., and Stent, G. S.,** Neuronal control of swimming in the medicinal leech. IV. Identification of a network of oscillatory interneurones, *J. Exp. Biol.,* 75, 25, 1978.
34. **Gardner-Medwin, A. R., Jansen, J. K. S., and Taxt, T.,** The "giant" axon of the leech, *Acta Physiol. Scand.,* 87, 30A, 1973.
35. **Gee, W.,** The behavior of leeches with especial reference to its modifiability, *Univ. Calif. Berkeley, Publ. Zool.,* 11, 197, 1913.
36. **Glover, J. C. and Kramer, A. P.,** Serotonin analog selectively ablates identified neurons in the leech embryo, *Science,* 216, 317, 1982.
37. **Granzow, B., Freed, E., and Kristan, B.,** Long-term behavioral and neuronal changes induced by section of interganglionic connections in leeches, *Soc. Neurosci. Abstr.,* 10, 686, 1980.
38. **Gray, J., Lissman, H. W., and Pumphery, R. J.,** The mechanism of locomotion in the leech *(Hirudo medicinals), J. Exp. Biol.,* 15, 408, 1938.
39. **Hagiwara, S. and Morita, H.,** Electrotonic transmission between two nerve cells in the leech ganglion, *J. Neurophysiol.,* 25, 721, 1962.
40. **Hartley, J. C.,** The life history of *Trochea subviridis, J. Anim. Ecol.,* 31, 519, 1962.
41. **Jansen, J. K. S. and Nicholls, J. G.,** Regeneration and changes in synaptic connectons between individual nerve cells in the central nervous system of the leech, *Proc. Natl. Acad. Sci. U.S.A.,* 69, 636, 1972.
42. **Johnson, J. E., Philpott, D. E., and Miquel, J.,** A study of axonal degeneration in the optic nerve of aging mice, *Age,* 2, 124, 1978.
43. **Keyser, K. T., Frazer, B. M., and Lent, C. M.,** Physiological and anatomical properties of Leydig cells in the segmental nervous system of the leech, *J. Comp. Physiol.,* 146, 379, 1982.
44. **Kisiel, M. J. and Zuckerman, B. M.,** Effects of centrophenoxine on the nematode *Caenorhabditis briggsae, Age,* 1, 17, 1978.
45. **Kramer, A. P.,** The nervous system of the glossiphoniid leech *Haementeria ghilianii.* II. Synaptic pathways controlling body wall shortening, *J. Comp. Physiol.,* 144, 449, 1981.
45a. **Kramer, A. P.,** personal communication.
46. **Kramer, A. P. and Goldman, J. R.,** The nervous system of the glossiphoniid leech *Haementeria ghilianii.* I. Identification of neurons, *J. Comp. Physiol.,* 144, 435, 1981.
47. **Kramer, A. P. and Kuwada, J. Y.,** Formation of the receptive fields of leech sensory neurons during embryonic development, *J. Neurosci.,* 3, 2474, 1983.
48. **Kretz, J. R., Stent, G. S., and Kristan, W. B., Jr.,** Photosensory input pathways in the medicinal leech, *J. Comp. Physiol.,* 106, 1, 1976.
49. **Kristan, W. B., Jr.,** Neuronal changes associated to behavioral changes in chronically isolated segments of the medicinal leech, *Brain Res.,* 167, 215, 1979.
50. **Kristan, W. B., Jr.,** The generation of motor patterns, in *Information Processing in the Nervous System,* Pinsker, H. and Willis, W. D., Eds., Raven Press, N.Y., 1980, 241.
51. **Kristan, W. B., Jr.,** Sensory and motor neurones responsible for the local bending response in leeches, *J. Exp. Biol.,* 96, 161, 1982.
52. **Kristan, W. B., Jr. and Calabrese, R. L.,** Rhythmic swimming activity in neurons of the isolated nerve cord of the leech, *J. Exp. Biol.,* 65, 643, 1976.
53. **Kristan, W. B., Jr. and Guthrie, P. B.,** Acquisition of swimming behavior in chronically isolated single segments of the leech, *Brain Res.,* 131, 191, 1977.
54. **Kristan, W. B., Jr., McGirr, S. J., and Simpson, G. V.,** Behavioral and mechanosensory neuron responses to skin stimulation in leeches, *J. Exp. Biol.,* 96, 143, 1982.
55. **Kristan, W. B., Jr., Stent, G. S., and Ort, C. A.,** Neuronal control of swimming in the medicinal leech. I. Dynamics of the swimming rhythm, *J. Comp. Physiol.,* 94, 97, 1974.
56. **Kristan, W. B., Jr., Stent, G. S., and Ort, C. A.,** Neuronal control of swimming in the medicinal leech. III. Impulse patterns of motor neurons, *J. Comp. Physiol.,* 94, 155, 1974.
57. **Kuffler, D. P.,** Neuromuscular transmission in longitudinal muscle of the leech *Hirudo medicinalis, J. Comp. Physiol.,* 124, 333, 1978.
58. **Kuffler, S. W. and Nicholls, J. G.,** The physiology of neuroglial cells, *Ergeb. Physiol. Biol. Chem. Exp. Pharmakol.,* 57, 1, 1966.
59. **Kuffler, S. W. and Potter, D. D.,** Glia in the leech nervous system: physiological properties and neuron-glia relationships, *J. Neurophysiol.,* 27, 290, 1964.
60. **Kulkarni, G. K., Nagabushanam, R., and Hanumante, M. M.,** Reproductive biology of the Indian freshwater leech *Poecilobdella viridis* (Blanchard), *Hydrobiologia,* 58, 157, 1978.

61. **Kuwada, J. Y. and Kramer, A. P.,** Embryonic development of the leech nervous system: primary axon outgrowth of identified neurons, *J. Neurosci.,* 3, 2098, 1983.
62. **Landfield, P. W. and Lynch, G.,** Impaired monosynaptic potentiation in *in vitro* hippocampal slies from aged, memory-deficient rats, *J. Gerontol.,* 32, 523, 1977.
63. **Landfield, P. W., McGaugh, J. L., and Lynch, G.,** Impaired synaptic potentiation processes in the hippocampus of aged, memory-deficient rats, *Brain Res.,* 150, 85, 1978.
64. **Landfield, P. W., Wurtz, C., and Lindsey, J. D.,** Quantification of synaptic vesicles in hippocampus of aging rats and initial studies of possible relations to neurophysiology, *Brain Res. Bull.,* 4, 757, 1979.
65. **Laverack, M. S.,** Mechanoreceptors, photoreceptors and rapid conduction pathways in the leech, *Hirudo medicinalis, J. Exp. Biol.,* 50, 129, 1969.
66. **Learner, M. A. and Potter, D. W. B.,** Life history and production of the leech *Helobdella stagnalis (L) (Hirudinea)* in a shallow eutrophic reservoir in South Wales, *J. Anim. Ecol.,* 43, 199, 1974.
67. **Lent, C. M.,** Retzius cells: neuroeffectors controlling mucus release by the leech, *Science,* 179, 693, 1973.
68. **Lent, C. M.,** Retzius cels from segmental ganglia of four species of leeches: comparative neuronal geometry, *Comp. Biochem. Physiol.,* A 44, 35, 1973.
69. **Lent, C. M.,** Fluorescent properties of monoamine neurons following glyoxylic acid treatment of intact leech ganglia, *Histochemistry,* 75, 77, 1982.
70. **Lent, C. M. and Frazer, B. M.,** Connectivity of the monoamine-containing neurones in central nervous system of leech, *Nature (London),* 266, 844, 1977.
71. **Lent, C. M., Ono, J., Keyser, K. T., and Karten, H.,** Identification of serotonin within vital-stained neurons from leech ganglia, *J. Neurochem.,* 32, 1559, 1979.
72. **Liss, R.,** Aging brain and dementia, in *Aging — its Chemistry,* Deitz, A. A., Ed., American Association of Clinical Chemistry, Washington, D.C., 1979, 183.
73. **Macagno, E. R.,** Number and distribution of neurons in the leech segmental ganglion, *J. Comp. Neurol.,* 190, 283, 1980.
74. **Magni, F. and Pellegrino, M.,** Neural mechanisms underlying the segmental and generalized cord short-ening reflexes in the leech, *J. Comp. Physiol.,* 124, 339 1978.
75. **Mann, K. H.,** The life history of *Erpobdella octoculata* (Linnaeus, 1758), *J. Anim. Ecol.,* 22, 199, 1953.
76. **Mann, K. H.,** A study of a population of the leech *Glossiphonia complanata* (L), *J. Anim. Ecol.,* 26, 99, 1957.
77. **Mann, K. H.,** The life history of the leech *Erpobdella testacea* Sav. and its adaptive significance, *Oikos,* 12, 164, 1961.
78. **Mann, K. H.,** *Leeches (Hirudinea). Their structure, physiology, ecology and embryology,* Pergamon Press, N.Y., 1962.
79. **Mason, A. and Kristan, W. B., Jr.,** Neuronal excitation, inhibition and modulation of leech longitudinal muscle, *J. Comp. Physiol.,* 146, 527, 1982.
80. **Miquel, J., Economos, A. C., and Bensch, K. J.,** Insect vs. mammalian aging, in *Aging and Cell Structure,* Johnson, J. E., Ed., Plenum Press, N.Y., 1981.
81. **Miquel, J., Lundgren, P. R., Bensch, K. J., and Atlan, H.,** Effect of temperature on the life span, vitality and fine structure of *Drosophila melanogaster, Mech. Aging Dev.,* 5, 347, 1976.
82. **Miyazaki, S., Nicholls, J. G., and Wallace, B. G.,** Modification and regeneration of synaptic connections in cultured leech ganglia, *Cold Spring Harbor Symp. Quant. Biol.,* 40, 483, 1976.
83. **Muller, K. J.,** Synapses between neurones in the centrol nervous system of the leech, *Biol. Rev.,* 54, 99, 1979.
84. **Muller, K. J. and Carbonetto, S.,** The morphological and physiological properties of a regenerating synapse in the C.N.S. of the leech, *J. Comp. Neurol.,* 185, 485, 1979.
85. **Muller, K. J. and Nicholls, J. G.,** Different properties of synapses between a single sensory neurone and two different motor cells in the leech C.N.S., *J. Physiol.,* 238, 357, 1974.
86. **Muller, K. J., Nicholls, J. G., and Stent, G. S., Eds.,** *Leech Neurobiology,* Cold Spring Harbor Laboratory, N.Y., 1981, 277.
87. **Muller, K. J. and Scott, S. A.,** Transmission at a ''direct'' electrical connexion mediated by an interneurone in the leech, *J. Physiol.,* 311, 565, 1981.
88. **Nicholls, J. G. and Baylor, D. A.,** Specific modalities and receptive fields of sensory neurons in the CNS of the leech, *J. Neurophysiol.,* 31, 740, 1968.
89. **Nicholls, J. G. and Purves, D.,** Monosynaptic chemical and electrical connexions between sensory and motor cells in the central nervous system of the leech, *J. Physiol.,* 209, 647, 1970.
90. **Nicholls, J. G. and Wallace, B. G.,** Modulation of transmission at an inhibitory synapse in the central nervous system of the leech, *J. Physiol.,* 281, 157, 1978.
91. **Nicholls, J. G. and Wallace, B. G.,** Quantal analysis of transmitter release at an inhibitory synapse in the C.N.S. of the leech, *J. Physiol.,* 281, 171, 1978.
92. **Nosal, G.,** Neuronal involution during aging, ultrastructural study in the rat cerebellum, *Mech. Aging Dev.,* 10, 292, 1979.

93. **Nusbaum, M. P. and Kristan, W. B., Jr.,** The swim-initiating ability of intersegmental serotonin-containing leech interneurons, *Soc. Neurosci. Abstr.,* 12, 161, 1982.

94. **Ort, C. A., Kristan, W. B., Jr., and Stent, G. S.,** Neuronal control of swimming in the medicinal leech. II. Identification and connections of motor neurons, *J. Comp. Physiol.,* 94, 121, 1974.

95. **Peng, M. T. and Lee, L. R.,** Regional differences of neuron loss of rat brain in old age, *Gerontologia,* 25, 205, 1979.

96. **Peretz, B. and Lukowiak, K.,** Age dependent CNS control of the habituating gill withdrawal reflex and of correlated activity in identified neurons in *Aplysia, J. Comp. Physiol.,* 103, 1, 1975.

97. **Peretz, B., Ringham, G., and Wilson, R.,** Age diminished neuronal function of central neuron L7 in *Aplysia, J. Neurobiol.,* 13, 141, 1982.

98. **Poon, M., Friesen, W. O., and Stent, G. S.,** Neuronal control of swimming in the medicinal leech. V. Connections between the oscillatory interneurons and the motor neurons, *J. Exp. Biol.,* 75, 43, 1978.

99. **Ready, D. and Nicholls, J.,** Identified neurones isolated from leech CNS make selective connections in culture, *Nature (London),* 281, 67, 1979.

100. **Richards, A. G. and Kolderie, M. Q.,** Variation in weight, development rate and hatchability of *Oncopeltus* eggs as a function of the mother's age, *Entomol. News,* 68, 57, 1957.

101. **Rogers, J., Silver, M. A., Shoumaker, W. J., and Bloom, F. E.,** Senescent changes in a neurobiological model system: cerebellar Perkinje cell electrophysiology and correlative anatomy, *Neurobiol. Aging,* 1, 3, 1980.

102. **Rogers, J., Zornetzer, S. F., Shoumaker, W. J., and Bloom, F. E.,** Electrophysiology of aging brain: senescent pathology of cerebellum, in *Brain Neurotransmitter and Receptors in Aging and Age-Related Disorders,* Enna, S. S., Samorajski, T., and Beer, B., Eds., Raven Press, N.Y., 1981, 81.

103. **Roth, G. S.,** Steroid and dopaminergic receptors in the aged brain, in *Brain Neurotransmitters and Receptors in Aging and Age-Related Disorders,* Enna, S. S., Samorajski, T., and Beer, B., Eds., Raven Press, N.Y., 1981, 163.

104. **Rude, S.,** Monoamine-containing neurons in the central nervous system and peripheral nerves of the leech *Hirudo medicinalis, J. Comp. Neurol.,* 136, 349, 1969.

105. **Samorajski, T.,** Normal and pathological aging of the brain, in *Aging 17: Brain Neurotransmitter and Receptors in Aging and Age-related Disorders,* Enna, S. S., Samorajski, T., and Beer, B., Eds., Raven Press, N.Y., 1981, 1.

106. **Sargent, P. B.,** Transmitters in the Leech Central Nervous System: Analysis of Sensory and Motor Cells, Doctoral dissertation, Harvard University, Cambridge, 1975.

107. **Sargent, P. B.,** Synthesis of acetylcholine by excitatory motoneurons in central nervous system of the leech, *J. Neurophysiol.,* 40, 453, 1977.

108. **Sawyer, R. T.,** Observations on the natural history and behavior of *Erpobdella punctata* (Leidy) (Annelida: Hirudinea), *Am. Midl. Nat.,* 83, 65, 1970.

109. **Sawyer, R. T.,** North American freshwater leeches, exclusive of the *Piscicolidae,* with a key to all species, University of Illinois Press, Urbana, 1972.

110. **Sawyer, R. T.,** Leech biology and behavior, in *Neurobiology of the Leech,* Muller, K. J., Nicholls, J. G., and Stent, G. S., Eds., Cold Spring Harbor Laboratory, N.Y., 1981, 7.

111. **Sawyer, R. T., Lepont, F., Stuart, D. K., and Kramer, A. P.,** Growth and reproduction of the giant glossiphoniid leech *Haementeria ghilianii, Biol. Bull.,* 160, 322, 1981.

112. **Schleip, W.,** Ontogenie der Hirudineen, in *Klassen und Ordnungen des Tierreichs,* Bronn, H. G., Ed., Vol. 4 (Part 2), Div. III, Book 4, Akad. Verlagsgesellschaft, Leipzig, 1936, 1.

113. **Smith, C. B., Goochee, C., Rapoport, S. I., and Sokoloff, L.,** Effects of aging on local rate of cerebral glucose utilization in the rat, *Brain,* 103, 351, 1980.

114. **Sohal, R. S.,** Relationship between metabolic rate, lipofuscin accumulation and lysosomal enzyme activity during aging in the adult housefly *Musca domestica, Exp. Gerontol.,* 16, 347, 1981.

115. **Sohal, R. S. and Sharma, S. P.,** Age related changes in the fine structure and number of neurones in the brain of the housefly *Musca domestica, Exp. Gerontol.,* 7, 243, 1972.

116. **Stent, G. S. and Kristan, W. B., Jr.,** Neural circuits mediating rhythmic movements, in *Leech Neurobiology,* Muller, K. J., Nicholls, J. G., and Stent, G. S., Eds., Cold Spring Harbor Laboratory, N.Y., 1981, 113.

117. **Stent, G. S., Kristan, W. B., Jr., Friesen, W. O., Ort, C. A., Poon, M., and Calabrese, R. L.,** Neuronal generation of the leech swimming movement, *Science,* 200, 1348, 1978.

118. **Stent, G. S., Thompson, W. J., and Calabrese, R. L.,** Neural control of heartbeat in the leech and in some other invertebrates, *Physiol. Rev.,* 59, 101, 1979.

119. **Stent, G. S. and Weisblat, D. A.,** The development of a simple nervous system, *Sci. Am.,* 246, 136, 1982.

120. **Stewart, W. W.,** Lucifer dyes — highly fluorescent dyes for biological tracing, *Nature (London),* 292, 17, 1981.

121. **Strehler, B. L.,** Introduction: aging and the human brain, in *Aging*, Vol.3, Terry, R. D. and Gershon, S., Eds., Raven Press, N.Y., 1976, 1.

122. **Stuart, A. E.,** Physiological and morphological properties of motoneurones in the central nervous system of the leech, *J. Physiol. (London)*, 209, 627, 1970.

123. **Thompson, W. J. and Stent, G. S.,** Neuronal control of heartbeat in the medicinal leech. I. Generation of the vascular constriction rhythm by heart motor neurons, *J. Comp. Physiol.*, 111, 261, 1976.

124. **Thompson, W. J. and Stent, G. S.,** Neuronal control of heartbeat in the medicinal leech. II. Intersegmental coordination of heart motor neuron activity by heart interneurons, *J. Comp. Physiol.*, 111, 281, 1976.

125. **Thompson, W. J. and Stent, G. S.,** Neuronal control of heartbeat in the medicinal leech. III. Synaptic relations of the heart interneurons, *J. Comp. Physiol.*, 111, 309, 1976.

126. **Tillman, D. L. and Barnes, J. R.,** The reproductive biology of the leech *Helobdella stagnalis* (L.) in Utah Lake, *Freshwater Biol.*, 3, 137, 1973.

127. **Van Damme, N.,** Organogénese de l'appareil genital chez la sangsue *Erpobdella octoculata* L. (Hirudinée-Pharyngobdelle), *Arch. Biol.*, 85, 373, 1974.

128. **Van Essen, D. C. and Jansen, J. K. S.,** The specificity of re-innervation by identified sensory and motor neurons in the leech, *J. Comp. Neurol.*, 171, 433, 1977.

129. **Wallace, B. G.,** Neurotransmitter chemistry, in *Leech Neurobiology*, Muller, K. J., Nicholls, J. G., and Stent, G. S., Eds., Cold Spring Harbor Laboratory, N.Y., 1981, 147.

130. **Weeks, J. C.,** Neuronal basis of leech swimming: separation of swim initiation, pattern generation and intersegmental coordination by selective lesions, *J. Neurophysiol.*, 45, 698, 1981.

131. **Weeks, J. C.,** Segmental specialization of a leech swim-initiating interneuron (cell 205), *J. Neurosci.*, 2, 972, 1982.

132. **Weeks, J. C.,** Synaptic basis of swim initiation in the leech. I. Connections of a swim-initiating neuron (cell 204) with motor neurons and pattern-generating "oscillator" neurons, *J. Comp. Physiol.*, 148, 253, 1982.

133. **Weeks, J. C.,** Synaptic basis of swim initiation in the leech. II. A pattern-generating neuron (cell 208) which mediates motor effects of swim-initiating neurons, *J. Comp. Physiol.*, 148, 265, 1982.

134. **Weeks, J. C. and Kristan, W. B., Jr.,** Initiation, maintenance and modulation of swimming in the medicinal leech by the activity of a single neuron, *J. Exp. Biol.*, 77, 71, 1978.

135. **Weisblat, D. A. and Blair, S. S.,** Developmental indeterminancy in embryos of the leech *Helobdella triserialis*, *Dev. Biol.*, 106, 326, 1984.

136. **Weisblat, D. A., Harper, G., Stent, G. S., and Sawyer, R. T.,** Embryonic cell lineages in the nervous system of the Glossiphoniid leech *Helobdella triserialis*, *Dev. Biol.*, 76, 58, 1980.

137. **Weisblat, D. A., Kim, S. Y., and Stent, G. S.,** Embryonic origin of cells in the development of the leech *Helobdella triserialis*, *Dev. Biol.*, 1984, in press.

138. **Weisblat, D. A., Sawyer, R. T., and Stent, G. S.,** Embryonic origin of cells by intracellular injection of a tracer enzyme, *Science*, 202, 1295, 1978.

139. **Weisblat, D. A., Zackson, S. L., Blair, S. S., and Young, J. D.,** Cell lineage analysis by intracellular injection of fluorescent tracers, *Science*, 209, 1538, 1980.

140. **Whitman, C. O.,** A contribution to the history of germ layers in *Clepsine*, *J. Morphol.*, 1, 105, 1887.

141. **Whitman, C. O.,** The embryology of *Clepsine*, *Q. J. Microsc. Sci.*, 18, 215, 1878.

142. **Wilkialis, J.,** Investigations on the biology of leeches of the *Glossiphoniidae* family, *Zool. Pol.*, 20, 29, 1970.

143. **Willard, A. L.,** Effects of serotonin on the generation of the motor pattern for swimming by the medicinal leech, *J. Neurosci.*, 1, 936, 1981.

144. **Willett, J. D., Rahim, I., Geist, M., and Zuckerman, B. M.,** Cyclic nucleotide exudation by nematodes and the effects on nematode growth, development and longevity, *Age*, 3, 82, 1980.

145. **Wilson, A. H. and Lent, C. M.,** Electrophysiology and anatomy of the large neuron pairs in the sub-esophageal ganglion of the leech, *Comp. Biochem. Physiol.*, 46A, 301, 1973.

146. **Wisniewski, H. M. and Terry, R. D.,** Neuropathology of the aging brain, in *Neurobiology of Aging*, Terry, R. D. and Gershon, S., Eds., Raven Press, N.Y., 1976, 265.

147. **Wrona, F. J., Davies, R. W., Linton, L., and Wilkialis, J.,** Competition and coexistence between *Glossiphonia complanata* and *Helobdella stagnalis* (Glossiphoniidae:Hirudinoidea), *Oecologia*, 48, 133, 1981.

148. **Yau, K. W.,** Physiological properties and receptive fields of mechanosensory neurones in the head ganglion of the leech: comparison with homologous cells in segmental ganglia, *J. Physiol.*, 263, 489, 1976.

149. **Young, S., Dedwylder, R., II, and Friesen, W.,** Responses of the medicinal leech to water waves, *J. Comp. Physiol.*, 114, 111, 1981.

150. **Zackson, S. L.,** Cell clones and segmentation in leech development, *Cell*, 31, 761, 1982.

151. **Zipser, B.,** Identifiable neurons controlling penile eversion in the leech, *J. Neurophysiol.*, 42, 455, 1979.

152. **Zipser, B.,** Identification of specific leech neurones immunoreactive to enkephalin, *Nature (London)*, 283, 857, 1980.

153. **Zipser, B. and McKay, R.,** Monoclonal antibodies distinguish identifiable neurones in the leech, *Nature (London),* 289, 549, 1981.

154. **Zuckerman, B. M. and Barrett, K. A.,** Effects of PCA and DMAE on the nematode *Caenorhabditis briggsae, Exp. Aging Res.,* 4, 133, 1978.

Chapter 6

# CELLULAR IMMUNITY IN EARTHWORMS AND POSSIBLE RELATIONSHIP TO AGING*

**Edwin L. Cooper**

## TABLE OF CONTENTS

*   This work was supported in part by grants BRSG (UCLA), AI 15976-03, NSF PCM 82-04879, and HD 09333-07.

# I. INTRODUCTION

## A. Overview

Invertebrates constitute the largest of all animal groups. They are versatile in their choices of food and life styles; therefore they are adaptable on land, in water, or in air. According to some estimations, there are more than 1 million species of invertebrates. They are a vital link in the food chain for humans and other animals.[1] In contrast, their sheer numbers also pose a threat to life since they themselves are vectors of certain diseases, or may be parasites and therefore the direct cause of disease. Aside from knowing how beneficial or harmful invertebrates may be, understanding more about how they live is valuable for heuristic reasons. As this chapter will point out, the invertebrates may soon become as popular in our quest for less expensive animal models for research on aging as they have in the past in, for example, genetics, embryology,[2] and immunology.[3]

## B. Invertebrates and Aging Research

Focusing on invertebrates as models for aging is important if we are to expand our knowledge of senescence as a biological phenomenon. In some instances, using invertebrates poses certain unique advantages when compared to other animals. First, the body plan of invertebrates is usually simpler and often smaller than that of most vertebrates. Second, they usually have shorter lifespans and are relatively easy to maintain, making them ideal for studies related to senescence. Third, invertebrates have been easily inbred, providing defined genetic stocks, another distinct advantage for their use in aging research. Finally, we are now forced to search for universal mechanisms that will serve to unify our concepts regarding aging; thus, information from all animals is necessary[4] (Table 1).[5] There may be interconnections between the immune system and aging as has been revealed in mammals, but this connection is not clear in invertebrates.[6] This chapter will review studies dealing with cell-mediated immunity and describe an uninvestigated degenerative syndrome in earthworms resembling senescence.

## C. Defense and Immunity in Lower Invertebrates

Among the one-celled animals (Protozoa) at least one class, the Sarcodina, has been important in studes of phagocytosis for demonstrating incompatibilities resulting from transplantation of nuclei between several different strains or clones.[1] Sponges (Porifera) are also of interest. They are the first metazoans, and their scavenging cells, the archeocytes, the most primitive macrophages, may be involved either directly by cell contact or indirectly by secretion of "killer" substances in the well-known incompatibility reactions now being actively studied.[7] It is at this point that we see the earliest evidence for the concept of *self/notself,* which is represented throughout the animal kingdom as fundamental to *recognition* which, in turn, is necessary for immune competence. Whether the first receptors for this *self/notself* distinction are found so early phylogenetically is unclear, but evolution surely has produced recognition molecules or receptors in every animal group.[8]

## D. General View of Taxonomy and Immune Characteristics

At the level of the coelenterates, it is believed that advanced immunoresponses diverged. The chordate line, to which vertebrates belong, consists of animals chracterized by the deuterostomate pattern of embryogenesis (i.e., two separate openings develop into the mouth and anus). The echinoderms and tunicates belong to this group. Among the protostomes, to which the annelids, arthropods, and molluscs belong, it is the blastopore alone which splits to form the mouth or anus during blastogenesis. This embryologic classification may be irrelevant if we deal with immunologic characteristics from which other taxonomic schemes could conceivably be developed.[9] With respect to current immunological data, some deuterostomes are surely more akin to the protostomes than to the vertebrates. Thus, there are immunological characteristics shared by all invertebrates.

Annelids have numerous leukocytic types with complex morphologies capable of affecting the archetypal defense reactions *phagocytosis* and *encapsulation.* Leukocytes are also capable of apparently more complex functions such as the recognition and destruction of foreign transplants as a cell-mediated reaction.[10] These leukocytes can also synthesize and secrete humoral substances which can lyse, agglutinate, or otherwise sequester foreign particulate antigens and thus provide the humoral immune defense system.[11]

## E. Early Studies on Senescence and Death in Invertebrates

Child[12] asserted that senescence does occur in invertebrates and is characterized by a decrease in metabolic rate with advancing age. According to Child, "senescence is a characteristic and necessary feature of life and occurs in all organisms." The causes of death were later speculated upon by Orton[13] and Szabo[14] who advanced the view that the main cause of physiological death from senescence may be traced to several organ systems and that this may vary with different species. Regarding earthworms, signs of approaching death were observed by Cameron.[15] These appeared as immobility, decreased responses to tactile stimuli, abnormal body attitude, i.e., lying on the lateral or dorsal surface rather than ventral;

**Table 1**
MAXIMUM AGES OF INDIVIDUALS IN VARIOUS PHYLA

| Phylum | Species | Max. age attained (years) | Evidence of age | References (see Ref. 4) |
|---|---|---|---|---|
| Porifera | *Suberites carnosus* | 15 | Captive individuals | Arndt |
| Coelenterata | *Cereus pedunculatus* | 85—90 | Captive individuals | Ashworth and Annandale |
| Flatworm | *Taeniorrhynchus saginatus* | 35+ | Individual case history | Penfold, Penfold and Phillips |
| Nematoda | *Wuchereria bancrofti* | 17 | Case history | Knabe |
| Annelida | *Sabella pavonia* | 10+ | Captive individual | Wilson |
| Arthropoda Arachnida | *Tarantula* | 11—20 | Kept in captivity | Baerg |
| Crustacea | *Homarus americana* (lobster) | 50 | Size of captured specimens | Herrick, 1911 |
| Insecta | *ermites* | 25—60 | Characteristics of wild specimens | Snyder |
| | *Stenamma westwoodi* (hymenoptera) | 16—18 | Captive individual | |
| Echinodermata | *Marthasterias glacialis* | 7+ | Captive individual | Wilson |
| Mollusca | *Venus mercenaria* | 40+ | Growth rings | Hopkins |
| | *Megalonaias gigantea* | 54 | Growth rings | Chamberlain |
| Rotifera | *Rotaria macrura* | 0.16 | Captive individual | Spemann |
| Vertebrates | | | | |
| Fish | *Silurus glanis* | 60+ | | Flower |
| Amphibia | Megalobratrachus | 52+ | | Flower |
| Reptiles | *Testudo sumeiri* | 152+ | | Flower |
| Birds | *Bubo bubo* | 68+ | | Flower |
| Mammals | *Homo sapiens* | 118+ | | Authenticated human birth and death records |

atonicity, the body being flabby or dry; localized constriction and dilation of segments; autotomy of terminal segments, and evagination of the pharynx and anus. Whether these changes preceding death are a natural consequence of senescence is not clear.

## II. THE OLIGOCHAETE ANNELIDS

### A. Characteristics

The oligochaetes, approximately 3000 species, as members of the phylum Annelida (ringed), are segmented protostomes with a well-developed blood-vascular system distinct from the coelom. The most familiar oligochaetes are the earthworms, which occupy a wide variety of habitats ranging from soil to fresh and salt water. Earthworms vary enormously in size, from less than 0.5 mm in members of the family Aeolosomatidae to over 3 m for the giant earthworm of Australia (Megoscolecidae).[16]

Since the bodies of earthworms consist of a series of rings or segments, the internal organs and the body walls are segmented. Thus, worms are composed of approximately 100 more or less similar units equipped with one or a pair of organs of each system. The segments are separated from each other by transverse septa. Earthworms are protected from desiccation by a thin, transparent cuticle secreted by cells of the epidermis which comprise the outer layer of the body wall. Additional protection is provided by glandular cells of the epidermis which secrete mucus. The body wall contains an outer layer of circular muscles and an inner layer of longitudinal muscles. Movement and anchoring are facilitated by four pairs of bristles or chaetae embedded in each segment except the first.

The digestive system consists of a pharynx, esophagus, and stomach divided into two parts, a thin-walled crop, a thick-walled muscular gizzard for grinding food, and thereafter, a long straight intestine for absorption — it terminates in the anus. The excretory system is comprised of paired nephridia, miniature kidneys consisting of a ciliated funnel, and a tube surrounded by a capillary network to facilitate removal of wastes from the bloodstream. The circulatory system consists of five pairs of muscular tubes or hearts, a major dorsal and ventral vessel, and numerous tributaries and capillary networks. The nervous system consists of a large, two-lobed aggregation of nerve cells, the brain located just above the pharynx in the third segment, and another ganglion just below the pharynx, the subpharyngeal ganglion in the fourth segment. A ring of nerves around the pharynx connects the two ganglia. From the lower ganglion, a nerve cord extends the entire length of the body ventrally beneath the digestive tract. In each segment there is a swelling of the nerve cord, a segmental ganglion from which nerves extend laterally to the muscles and organs. Although hermaphroditic, earthworms must copulate to ensure cross-fertilization. Eggs are protected in a tough cocoon from which emerge several young miniature worms after a period of complex embryonic developmental stages.

## B. The Coelom and the Immune System

Within the segmented coelom, the fluid acts as a hydraulic skeleton and also contains a number of free cells, or coelomocytes, which comprise the worm's immune system. The coelomic fluid in which these cells are bathed constitutes: (1) the humoral component of the immune system; and (2) an intermediate in transporting gases, food, and wastes between the circulatory system and individual cells. The blood vascular system is separate from the coelom and may contain a distinct cell population, about which relatively little is known when compared to coelomocytes.

Detailed information on structure and function of coelomocytes is available for only a few families, primarily the Lumbricidae, Megascolecidae, and Enchytraeidae. Coelomocytes of several species of Lumbricidae have recently been studied extensively in regard to their role(s) in cell-mediated immunity,[17-23] and will be reviewed here. The complex morphology of coelomocytes, emphasizing the oligochaetes, has been recently reviewed.[10]

## C. Origin and Formation of Coelomocytes

Coelomocytes are probably derived from the epithelial lining of the coelomic cavity or from specialized structures associated with the epithelium. The coelomic epithelium has been modified to form specialized "lymphoid organs" considered by some investigators to be leukopoietic in many species.[24] These "lymphoid organs" are segmentally arranged, paired nodules located throughout the intestinal region on either side of the dorsal vessel and attached to the anterior face of the septa. Different coelomocyte types, as well as cells with a lymphocyte-like morphology considered by Kindred[24] to be stem cells, are found in the "lymphoid organs." However, there has been disagreement concerning the presence of mitotic figures.[10]

By introducing foreign substances into the coelomic cavity, coelomocyte numbers will increase. Thus, coelomocytes are apparently derived directly from the coelomic epithelium in most species since it proliferates in response to these substances.[25] Coelomocytes of Lumbricidae consist of two primary cell lines: amoebocytes and eleocytes. Amoebocytes are derived from the parietal and septal epithelia; eleocytes from the epithelium covering the viscera and blood vessels.[15,26] In general, it is the amoebocyte which figures prominently in earthworm cell-mediated responses.

## III. WHAT IS KNOWN ABOUT THE EARTHWORM'S IMMUNE SYSTEM?

### A. Introduction

Earthworms bring to mind fishing, mass exit of worms from earthly caverns after spring rains, and the improvement of soil by their constant churning and digging, as Darwin observed.[27] Some view worms as ecologically sound agents for destroying wastes. They could be rich sources of edible protein during food shortages. Comparative immunologists interested in the phylogenesis of immunity have now rediscovered the earthworm[28-33] in a search for the evolution of man's complex immune system.[34,35] Animals like worms may possess stem cells with certain features from which higher immune systems evolved (Figure 1).

The immune systems of animals in the phylum Annelida (e.g., the earthworm) possess one important characteristic, i.e., they have developed a coelom containing coelomic fluid in which coelomocytes, the worm's leukocytes, are suspended. Coelomocytes, like leukocytes, are sensitive to perturbations such as infections and are active in defense reactions ranging from phagocytosis to the more complex mechanisms of tissue graft rejection. To react against foreign material, coelomocytes surely possess cellular recognition units or receptors whose nature is unknown. The receptors for antigens in mammals are antibodies but antibodies have not been found in earthworms. Worms must, therefore, have different, but not necessarily simpler, primordial receptor units. The fundamental problem in studies on the earthworm's immune system, as well as that of other invertebrates, is to explain the process of recognition.

### B. General Features of the Earthworm's Immune System

The crucial work of Metchnikoff and other early comparative immunologists working with invertebrates (including earthworms) led to knowledge of cellular immunity. Metchnikoff's contributions to immunology served to divide the field from its original emphasis on humoral immunity into its second major subdiscipline concerned with cellular immunity, which began by emphasizing the role of phagocytosis of foreign antigens, particularly infectious microorganisms. During the early 1960s, these two chief facets of the immune response, humoral and cellular immunity, were "unified" by revealing that each resulted primarily from products of two groups of lymphocytes. B-cells, when stimulated, will synthesize and secrete antibody, whereas T-cells help B-cells, assist in regulating the responses of B and other cells, and mediate cellular reactions such as graft rejection and immunity against cancer.

Some comparative immunologists have assumed that the earthworm's graft rejection capacities are mediated by T-cell progenitors, whereas others have disagreed with this point.[36,37] Phagocytosis is the defense reaction common to all animals, even the simplest, the unicellular protozoans. Earthworms contain a multitude of coelomocytes patrolling the coelomic cavity in search of foreign cells or substances. Since the end of the 19th century, we have known that any material introduced into the coelomic cavity will be phagocytosed and destroyed by coelomocytes. Earthworms are therefore no different from other animals in possessing an efficient, generalized, but nonspecific mechanism for disposing of foreign, *non-self* material. Nonspecific immune responses are universal and in the earthworm they may be even more efficient than equivalent responses in man.

In contrast, the specific immune system of earthworms possesses certain characteristics common to the immune systems of vertebrates, but it may be less complex.[38] There are antigen-recognizing cells and effector cells. Antigen-recognizing cells may be able to sense antigen via the as-yet-undemonstrated cell surface receptors. It is assumed that these putative antigen-recognition cells later interact with effector cells, a process which is responsible for at least one immune function, the rejection of transplants.

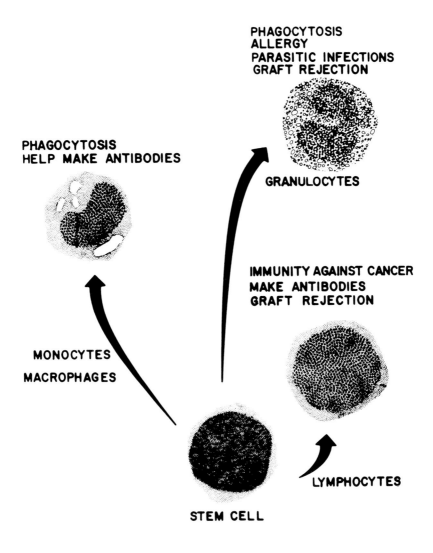

PHAGOCYTOSIS
ALLERGY
PARASITIC INFECTIONS
GRAFT REJECTION

GRANULOCYTES

PHAGOCYTOSIS
HELP MAKE ANTIBODIES

IMMUNITY AGAINST CANCER
MAKE ANTIBODIES
GRAFT REJECTION

MONOCYTES
MACROPHAGES

LYMPHOCYTES

STEM CELL

FIGURE 1.   Cells of the vertebrate immune system and their role in cellular and humoral immunity.

## 1. Graft Rejection

How are worms capable of distinguishing *self*, which is nonantigenic, from *nonself*, which is antigenic and potentially harmful or infectious? Let us deal with how worms reject grafts as a paradigm. First, worms will accept autografts permanently after initial healing and show no signs of rejection. However, allogeneic and xenogeneic transplants from other earthworms will first heal in, remain temporarily unaffected, rest unchallenged, but then later, most of these transplants will show gross and histological signs of graft rejection. The grafts are destroyed by coelomocytes, among which there are surely the effectors in addition to those which first recognized the antigen. There is no direct evidence that antigen-recognizing cells are concerned with graft rejection, but antigen-recognizing coelomocytes can bind to certain antigens found on sheep erythrocytes to form rosettes. Although presumably not related to the antigens on grafts, the adherence of sheep erythrocytes to coelomocytes suggests that receptors are present.

*2. Assays for Weak "Memory" against Transplants*

Graft rejection is presumably mediated by primed coelomocytes that infiltrate the graft matrix and once inside, attack viable muscle by phagocytosis. Electron micrographs reveal the destruction of viable graft musculature and support the view that the rejection process is not initiated against dead transplants. Moreover, grafts are always vascularized; thus, they are nourished and some often show signs of irritability due to innervation. That the mechanism of graft rejection possesses a degree of specificity has been demonstrated by several experiments: (1) accelerated rejection of second grafts; (2) adoptive transfer of the response by coelomocytes; (3) specificity of local responses; and (4) specificity of third-party rejection. In other words, there is evidence that primed, sensitized coelomocytes are capable of remembering a prior encounter with antigen.[39]

## C. Aspects of the Humoral Immune System

We are sure that like other invertebrates, worms possess mechanisms for ridding themselves of soluble and particulate material, chiefly by means of humoral components in the coelomic fluid such as agglutinins, lysins, and other bactericidal substances (Figure 2). Earthworms probably synthesize these components both in response to new antigenic challenges and to remove effete components. The chemical structure of these humoral components is not clear, but assuming they are akin to those found in other invertebrates such as oysters, they are probably specialized proteins.[40-43] The most unusual and interesting aspect of these components is that the agglutinins in particular, may be synthesized by coelomocytes and then released into the coelomic fluid. One model for the generation of receptors is that receptors bound to the surface of coelomocytes, when brought into contact with specific antigens, could be stimulated to synthesize more receptor, which is then shed into the coelomic fluid.

## D. A Summary Comment on Evolution and Immunity

Earthworms are important to immunology; their recognition and rejection of transplants and their synthesis of humoral components provide tangible and repeatable evidence that certain features of the immune response are common to all living creatures and that to reach its present, more complicated form in man, the immune response required a long evolutionary history.[44] In contrast, existence of the earthworm's immune system suggests that immune responses of higher species are not essential since the earthworm's survival is ensured by what are apparently less complicated immune capabilities.

## IV. TRANSPLANTATION OF AUTOGRAFTS, ALLOGRAFTS, AND XENOGRAFTS

## A. Accelerated Rejection of Intrafamilial Xenografts and Allografts. Is this Memory?

Earthworms have been important in studies concerning analyses of cell-mediated immunity since they can recognize and reject foreign tissue grafts, but accept grafts of self-tissue. All grafts, both foreign (allografts, xenografts) and self, heal initially and become vascularized and firmly attached to the host, but foreign grafts later show varying degrees of rejection, whereas self-tissue grafts (autografts) remain permanently healed. Xenografts are rejected more vigorously and usually completely, whereas allografts are rejected more slowly and often incompletely.[45-58]

Second-set *Eisenia foetida* xenografts on *Lumbricus terrestris* hosts undergo accelerated rejection if transplanted 1 week after grafting first-set xenografts. In contrast, second-set xenografts transplanted at longer intervals after first-set grafting, do not show accelerated

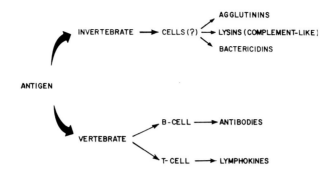

FIGURE 2.   Humoral and related cell immune responses in invertebrates and vertebrates.

rejection. The essential variable in these experiments has been the time of second-set transplantation. Accelerated rejection as a measure of weak or short-term memory is important in earthworms and other invertebrates and probably results from intense responses (although of short duration) of activated coelomocytes.[59] Accelerated rejection of *allo*transplants under these kinetic conditions agrees with earlier intrafamilial xenotransplant results obtained in earthworms, suggesting that an essential experimental variable is the time of second-set transplantation.[60]

## B. Are Coelomocytes Involved in the Destruction of Grafts?
### *1. Transplant Destruction: Histological Features*

Within 24 hr after transplanting grafts, coelomocytes first accumulate near the graft sites and infiltrate the graft matrix in an apparently nonspecific reaction to injury. By comparison, greater numbers of coelomocytes accumulate under, as well as within, xenografts than in autografts,[48,49,56] continuing to invade xenografts and remaining until destruction is complete. Coelomocytes can also surround xenografts in a second kind of reaction, an encapsulation-like process that may vary according to the extent of injury to the graft-bed of the host by the surgical procedures.[56,57]

In addition to light microscopy, xenograft rejection (*E. foetida* grafts on *L.terrestris* hosts) has also been studied by electron microscopy. Within 1 to 3 days posttransplantation, the graft matrix is infiltrated by coelomocytes known as type-I granulocytes (neutrophils, by light microscopy). Second, by 3 days posttransplantation, destruction of the inner muscle layer results in disorganized muscle fibers leaving large vacuolated areas. Type-I granulocytes are numerous and contain large phagocytic vacuoles with fragments of apparently viable muscle fiber. Third, at day 5, type-I granulocytes contain numerous phagosomes, residual bodies, and glycogen granules. Finally, by 11 to 13 days, type-I granulocytes have migrated into the outer layer of muscle which has almost been destroyed, leaving the outer epithelium mostly intact. Small lymphocytic (basophilic) coelomocytes are occasionally visible within the graft however, their function appears to be primarily that of minor scavenger, phagocytosing small cellular debris but not viable muscle fragments.

During graft destruction, muscle damage appears to be progressive, i.e., it occurs earliest in muscle layers closest to the coelomic cavity. During the further migration of coelomocytes into the outer muscle layers, destruction becomes evident in these areas. Coelomocytes infiltrate autografts and the wounded area of sham-operated worms in a nonspecific, general inflammatory response, related to wound-healing and removal of damaged tissue, with no visible phagocytosis of viable muscle.[58] Electron microscopic studies by Valembois[61,62] are somewhat different in xenograft experiments, employing *E. foetida* as host. In *E. foetida,* destruction of first-set grafts apparently occurs as a result of autolysis, involving synthesis of lysosomal enzymes by the transplanted tissues. In contrast, destruction of second-set grafts results from coelomocyte invasion, as in first-set xenografts of *L. terrestris*. Valembois has identified the coelomocytes which infiltrate the graft as macrophages (type-I granulocytes) that have their origin in stem cells of the splanchnopleural epithelium.

### 2. Kinetics of Coelomocyte Responses to Transplants

Grafting or injury to the earthworm's body wall is accompanied by an increase in free coelomocyte numbers beneath the grafted or wounded area. Hostetter and Cooper[63] have quantitatively correlated this response to graft rejection and wound-healing by measuring coelomocyte numbers directly under autografts, xenografts, and wounds at different times during rejection (or, in the case of autografts or wounds, healing) using *E. foetida* on *L. terrestris* hosts. Wounds induce rapid responses as coelomocytes rise within 24 hr, decline rapidly, and return to normal by 72 hr. Autografts elicit a weaker response, with coelomocyte counts rising only slightly above those of ungrafted worms and returning to normal by day 3. Coelomocyte responses to xenografts is slower, reaching a peak in 3 to 4 days; the decline is also slower, requiring 7 days for coelomocytes to return to normal levels.

In these experiments, xenograft destruction was complete in a mean time of 17 days at 20°C, approximately 10 days following the decline in coelomocyte numbers. Transplanted second grafts at this time (17 days posttransplantation of first xenograft) were rejected in an accelerated mean time of 6 to 7 days. Quantities of coelomocytes associated with second grafts or second wounds varied considerably from those associated with first grafts or wounds. After second xenografts, coelomocyte numbers were 20 to 30% higher than after first grafts, peaking 1 to 2 days posttransplantation. These accelerated and heightened responses against second grafts suggest that a memory component is inducible in earthworm coelomocytes.

### 3. Short-term Memory Transferred Adoptively

Coelomocyte involvement in graft rejection is based upon additional experimental evidence demonstrating adoptive transfer.[46,64,65] Host *L. terrestris* (A) have been first xenografted with *E. foetida,* then coelomocytes harvested at 5 days posttransplantation, and injected into ungrafted *L. terrestris* (B). These second hosts were then xenografted with the same *E. foetida* donor used to induce immunity in A. Because *L. terrestris* shows only negligible allograft responses late after transplantation,[48] no early coelomocyte allo-incompatibility was expected prior to the action of primed coelomocytes against the *E. foetida* graft transplanted to a *L. terrestris* host (B). The second host (B) showed accelerated rejection of its first transplant, demonstrating short-term memory and confirming that primed coelomocytes can adoptively transfer graft rejection responses. Adoptive transfer is cell-mediated since transfer of coelomic fluid alone, free of coelomocytes, is ineffective. Coelomocytes from unprimed *L. terrestris* or from *L. terrestris* primed with saline are also unable to transfer the response. How specific this adoptive transfer response is remains to be determined.

## C. Response of Coelomocytes to Tissue Transplantation Antigens In Vivo

In addition to mitogens, coelomocytes are stimulated by putative transplantation antigens. Lemmi[66] demonstrated the in vivo synthesis of DNA, measured by $^3$H-thymidine incorporation into coelomocytes of xenografted *L. terrestris* at various times postgrafting, with a peak incorporation at 4 days. Roch et al.[67] investigated in vitro DNA synthesis by *E. foetida* coelomocytes following wounding, allografting, and xenografting. Peak incorporation of $^3$H-thymidine occurred 4 days after all three operations, but was greatest after allografting (with an estimated stimulation index [S.I.] of approximately 15), less after xenografting (S.I. $\cong$ 5), and least after wounding (S.I. $\cong$ 4). Responses following second grafts or wounds were more complex, with a peak at day 2 for xenografts (S.I. $\cong$ 4), at days 2 and 6 for allografts (S.I. $\cong$ 4), and at day 6 for wounds (S.I. $\cong$ 3). Xenografting with the use of autoradiographic techniques (particularly second sets) stimulated DNA synthesis in greater numbers of small coelomocytes ($<10$ $\mu$m) than in larger cells ($>10$ $\mu$m).

Regarding DNA synthesis, stimulation of coelomocytes by both wounding and grafting is greater than that induced by mitogens.[68] Coelomocytes (13%) incorporated $^3$H-thymidine in vitro 4 days after xenografting; after second xenografts, less (6%) were labeled. After first wounds, 6.5% were labeled, and after second wounds, only 4%. Small cells appeared to be stimulated more by grafting than by wounding. In cells not stimulated by grafts or wounds, larger coelomocytes characterized by abundant cytoplasm, a few free ribosomes, some polyribosomes, moderate amounts of endoplasmic reticulum and vesicular elements, and large pseudopods incorporated $^3$H-thymidine more frequently than did small coelomocytes characterized by a by a high nucleo-cytoplasmic ratio and many free ribosomes. After second xenografts, however, the situation is reversed, with more small cells labeled than large cells. Coelomocytes do respond to mitogens, wounds, and grafts — responses that show some degree of specificity. This new information adds further support to existing evidence for the earthworm's primitive immunologic capacity.

## D. Preferential Migration of Coelomocytes In Vitro

Coelomocytes of *L. terrestris* were found by Marks et al.[69] to respond in vitro by directional migration to both bacterial and foreign-tissue antigens. With respect to bacteria (*Aeromonas hydrophila* and *Staphylococcus epidermidis,*) the magnitude of chemotactic responses was directly proportional to the bacterial concentration. Responses to body-wall tissues from *E. foetida*, *Pheretima sp.*, and *Tenebrio molitor* (an arthropod) were highest toward *E. foetida*, moderate toward *Pheretima sp.*, and least toward *T. molitor,* an inverse ratio to phylogenetic relatedness. According to their interpretation, the *Eisenia* chemotactant is a small molecular-weight protein whose activity is dialyzable and destroyed by heat. A factor which suppresses migration appeared to be present in *L. terrestris* body-wall tissue, since placing both *E. foetida* and *L. terrestris* tissue together in the chemotactant chamber resulted in reduced migration from that found with *E. foetida* tissue alone. The responding coelomocytes (92 to 94%) were neutrophils (Type-I granulocytes).

## V. ORIGIN OF THE MAJOR HISTOCOMPATIBILITY COMPLEX (MHC)

## A. Introduction

Studies on graft rejection are crucial for several reasons. First, the second-level immune response capacity, the T-cell type reaction, is demonstrable throughout the animal kingdom as an aspect of cell-mediated immunity. Second, and perhaps more important, the results of these simple experiments on graft rejection can be interpreted in a larger immunologic context. The data from these studies suggest that the genes controlling graft-rejection reactions also are involved in the capacity to synthesize immunoglobulins, the third level immune response capacity, and an aspect of humoral immunity relegated to B cells in higher animals. These histological incompatibilities have been the subject of several reviews; however, Klein's[70] speculations on the evolution of the major histocompatibility complex have been used most extensively, and to some extent, those by Cohen and Collins[71] and duPasquier.[72]

## B. General Considerations
### 1. Definitions

The major histocompatibility complex (MHC) is a cluster of genetic loci occupying a single chromosomal region whose products govern certain immunological functions.[70] These functions include induction of B-cell differentiation leading to humoral antibody production detectable by standard serological methods, induction of T-cell differentiation resulting in blast transformation, and the production of lymphocytes known as cytotoxic effector cells capable of killing target cells. By mixing lymphocytes in vitro, or by observing graft-versus-host reactions (GVH) in vivo, blast transformation can be detected. In testing for these effector T-cells, immunologists look for cell-mediated lymphocytotoxicity (CML) in vitro, GVH reactions, rejection of allografts, and delayed-type hypersensitivity (DTH) reactions in vivo. The MHC locus governs whether immune responses will be either high or low during humoral antibody production or cellular responses such as delayed-type hypersensitivity reactions, allograft rejections, or T-cell proliferation in vitro. Finally, complement biosynthesis and activation are also controlled by the MHC.

## 2. The MHC Divisions

The MHC can be divided into regions, and these in turn into classes. An MHC region is that portion of the complex delimited by at least one genetic marker separable by genetic crossing-over from other loci. Each region is presumably capable of coding for a single polypeptide chain. Regions are often divided into subregions — portions of the MHC identified by distinct but functionally related marker loci and, therefore, separated by crossing over. Klein[70] uses class for the first time to designate regions related by their genetic origin and/or function. Three different classes are designated (I, II, III) based upon differences in biochemical properties, phenotypic expression and function. For relevance to this chapter, we will deal only with Class I regions.

## C. Class I

### 1. General Characteristics

Class I regions occur in pairs with the two members of each pair closely linked but separated by other loci. During their evolution, Schreffler et al.[73] suggest, members of a pair originated from an ancestral gene by gene duplication. The products of Class I regions are glycoproteins whose monomers have a molecular weight of 45,000 daltons. They are noncovalently associated with a single chain of $\beta_2$-microglobulin. Class I antigens are expressed in the plasma membrane of all cells except some cells and some neoplasias. Mature vertebrate T- and B-lymphocytes and macrophages possess the highest concentration of Class I products, and whenever in allogeneic combinations, Class I molecules will induce the production of humoral antibodies detectable by serological methods. It is not known whether invertebrate blood cells possess Class I antigens.

### 2. The MHC in Invertebrates

### a. Evolution of MHC

An explanation for MHC evolution is difficult for a number of important reasons. First, immunology, unlike other disciplines, began with humans, and is only recently focusing more on the evolution of mechanisms by analyzing more primitive systems. Second, interpretations are often difficult since comparative immunologists have usually been educated first as biologists; by contrast, mammalian immunologists are highly specialized, usually much less versed in biology. Despite these problems, we are sure that histocompatibility reactions are demonstrable in all metazoan invertebrates even though there is still more information to be gained, especially concerning mechanisms.

### b. Summary of Histocompatibility Reactions in Invertebrates

Histocompatibility reactions in invertebrates are demonstrable in multicellular organisms, although the term cytoincompatibility can be used for unicellular protozoans.[74] Recognition of nonself has been shown to occur in sponge cells (specific cell aggregation) of the phylum Porifera and in embryonic tissues, recalling the original definition that Class I antigens are expressed in varying degrees on all plasma membranes. Acute, aggressive reactions have been described in Cnidaria and encapsulation in Arthropoda and Mollusca. These histocompatibility reactions occur when tissues of two genetically different individuals come into contact and fusion is prevented. In contrast, there are histocompatibility reactions that result from slow aggression of one tissue against another — also first demonstrated in sponges and coelenterates — but the genetic controls have been studied more elaborately by histoincompatibility reactions described in coelenterates *(Hydractinia echinata,* Hydra, and Gorgonians). Such histoincompatibility reactions require more study for understanding evolutionary relevance to allograft reactions and the existence of immunologic specificity and memory.

Allograft reactions (primitive) have been studied most extensively in advanced invertebrates, the Annelida (earthworms), Echinodermata (sea stars), and the Tunicata (sea squirts). Recently, new results have emerged from studies of these groups and a less primitive one, the ribbon worms (Nemertea). Essentially, the rejection process in all groups is mediated by leukocytes, transferable adoptively to nonsensitized recipients and characterized by a memory component. This is consistent with the suggestion that graft rejection is controlled by a histocompatibility locus. There are no reports on genetic control of allograft reactions in the Echinodermata, nor its transfer by adoptive means, although the full array of cells corresponding to vertebrate leukocytes is involved. Incompatibility reactions in Tunicata are not clearly explained, although fusion or lack of fusion between various colonies of the genus *Botryllus* may be controlled by a single genetic system with multiple genes.

Class I molecules stimulate strong effector T-cell production such as acute allograft reactions in vivo and cell-mediated lysis in vitro — the characteristics of Class I regions studied extensively from the phylogenetic perspective. The results from this approach to the MHC strongly support the argument offered by Ohno[75] that the gene(s) controlling any reactions evolved from an earlier gene. The origin of the primordial gene can be traced through phylogeny by analyzing progressively more primitive species.

## D. Occurrence, Evolution, and Function of MHC
### 1. Preservation of Allogeneic Individuality
Allogeneic incompatibility is demonstrable in all animals; thus, there should be evidence of the ancestral MHC. Why did animals evolve the capacity to react against foreign cells, especially in allogeneic situations, since in vertebrates and advanced invertebrates, the experimental confrontation of a host with a foreign allograft is an unnatural situation? The only instance in vertebrates in which allogeneic confrontation is natural is the fetus during pregnancy and neoplasia. It is in primitive species, however such as sponges, corals, and sea anemones, where opportunity for body contact is great and loss of individuality (or self) by tissue fusion is easy and of real danger by species (inter, intra) encroachment. Thus, invertebrates probably evolved first a recognition system that protects one species against uncontrolled fusion with another individual. In more advanced invertebrates and vertebrates, the external threat to loss of species individuality is minimized due to lack of colonial growth. However, the increase in parasites and in malignant growth within an organism may have been the evolutionary pressure that caused the development of a surveillance system that patrols, so to speak, the internal environment. The first stage in the evolution of a surveillance system was a recognition system with specific receptors; lymphocytes and a specific memory component developed later.

### 2. Relationship to Immunological Evolution
Responses are probably associated with precursors of T-lymphocytes in invertebrates, functional T-lymphocytes in ectothermic vertebrates, and true T-lymphocytes in mammals and birds; membrane-fine structure has been more fully characterized in mammals. The MHC and immunoglobulin (Ig) loci may be evolutionarily related. Class I molecules have been shown to be noncovalently associated with $\beta_2$-microglobulin molecules, and $\beta_2$-mi-

croglobulin has some homology with the Fc region. Second, protein A derived from the bacterium *Staphylococcus aureus* is known to bind to the Fc region of IgG and with H-2 antigen Finally, Class I and Ig molecules have been claimed to have similar tertiary structure, i.e., each molecule is composed of the two light chains and two heavy chains Considerable controversy remains around the homology, or lack of it, in Class I and Ig molecules resulting from amino-acid sequence analyses. If Class I and Ig loci are shown to have evolved from common ancestors, we can assume that the phylogenetic separation of its various branches probably occurred quite early in phylogeny. These emerging conclusions would serve to unify immunology via the MHC, which evolved first in primitive animals, perhaps in association with their capacity to show allogeneic incompatibility.

### E. $\beta_2$-Microglobulin-like Molecules in Invertebrates

$\beta_2$-Microglobulin ($\beta_2$-m), present on the surface of lymphocytes and many other body cells, is a single polpeptide chain of 99 amino-acid residues first isolated from human urine.[76] The precise function of $\beta_2$-m in mammals remains unclear, although it is a protein noncovalently associated with MHC heavy chains (HLA or H-2) on the plasma membrane.[77] The primary sequence of the human $\beta_2$-m shows a high degree of homology with the $C_H3$ domain of human IgG1,[78] suggesting its origin from a common ancestral gene for Ig and Class I histocompatibility antigens involved in organ transplantation.

By using radioimmunoassays, Shalev et al.[79] recently demonstrated $\beta_2$-m-like molecules in total extracts of earthworms and several other invertebrates. Their analysis did not involve, however, a search for $\beta_2$-m in association with cell membranes. Recently, Roch and Cooper[80] demonstrated $\beta_2$-m-like molecules on leukocyte membranes of the earthworm *Lumbricus terrestris* using polyclonal and monoclonal anti-human $\beta_2$-m antibodies. Only one cell type, the acidophil, which constitutes about 15% of the total leukocyte fraction was labeled. There may be no correlation between $\beta_2$-m and histocompatibility antigens in earthworms, unlike mammals, where such antigens are involved in organ transplantation.

## VI. DEGENERATIVE CHANGES IN EARTHWORMS SUGGESTIVE OF SENESCENT CHANGES

### A. Introduction

The early studies dealing with aging in annelids have been restricted to longevity records,[4,5,81] whereas descriptions of tissue and organ changes are lacking. Hylton[82] described aging changes in the flight muscle of mosquitoes; one change common to this observation and ours in annelids is the loss of muscle fibers. During studies on transplantation immunity in earthworms (*L. terrestris*), we became intrigued by the marked wasting and death of many allo- and xenografted worms, usually after the 20th week in the laboratory. These degenerative changes did not seem to be due to immunobiological experiments associated with transplantation.[83] In this degenerative syndrome, the body wall and organ systems are conspicuously affected, mainly by muscular atrophy, fibrosis, and deposits of congo red-positive, hyalinized substances predominantly in the nephridia and large blood vessels.

### B. Gross and Histopathologic Observations
*1. Gross Changes*

This degenerative syndrome appeared grossly as profound weight loss, shortening of body length, and apparent melanosis, maybe due to the presence of the aging pigment lipofucsin. Contents of the coelomic cavity were easier to visualize in degenerate worms than in new ones even though the body walls seemed darker. Experimental worms (long residents in the laboratory) were sluggish and hypo- or insensitive to light and other physical stimuli, no doubt a natural consequence of cachexia and deterioration of neural elements of the epidermis and nervous system. New worms from the wild always varied in length from 5 to 8 in, while those housed in the laboratory for 26 weeks measured 4 to 6 in. Just after the first week, the initial weight increase was evident, but this decreased to at least half of the initial value by the end of 29 weeks.

## 2. Histologic Changes

**Body wall** — Muscle atrophy and fibrosis were the principal histologic changes. Normal body wall consists of pseudostratified cuticular epithelium 19 to 21 μm thick with four major cell types: serous, mucous-secreting, columnar (supporting), and basal. Epidermal sense organs composed of neuroepithelial, basal, and supporting cells are found in some parts of the epidermis interspersed among the other epithelial elements. Muscle layers 100 to 400 μm thick are arranged into outer circular and inner longitudinal layers. Connective-tissue cells and fibers were sparse; only a few fine, aniline, blue-staining, and PAS-positive fibers were present between muscle bundles, but substantially increased at the junction of the two muscle layers where numerous circumferentially oriented blood vessels were found. Simple squamous epithelium lined the free surface of the internal muscle layer, in effect providing a barrier to the coelomic cavity (Figure 3C). The body walls of degenerate worms, by contrast, showed many histopatholoic changes (Figure 4C). The epithelium was only 8 to 9 μm thick; all cells were similar with acidophilic, cytoplasmic granules and well-defined nuclei with clear, large nucleoli. Each muscle layer was about 16 to 17 μm thick and only a few normal circular muscle fibers were observed beneath the basement membrane. Much more connective tissue was present, resulting in a connective tissue: muscle ratio of 4:1. The longitudinal muscle was very much reduced, but connective tissue elements were much less conspicuous than in controls. Fewer blood vessels were discerned, and pigment cells seemed to have increased. Many pyknotic nuclei characterized the peritoneum.

**Digestive system** — Chloragogen tissue in experimental worms was greatly thickened to 80 μm from the normal 20-μm thickness. Cell boundaries were not clear, nuclei were pyknotic, and the nuclear membranes thick and wrinkled. The cytoplasm was crowded with larger, membrane-bound granules that stain with either Masson or hematoxylin and eosin. The spirally arranged muscle layer between the chloragogen tissue and epithelium was reduced from 7 to 12 μm to 5 μm, while the basement membrane beneath the epithelium was thickened to twice or three times normal and highly scalloped. The intestinal epithelium thinned out from 40 μm to 21 to 28 μm pseudostratified with microvilli; the basal two thirds of the cells were vacuolated, and the nuclei were pyknotic (Figures 3D, 4D).

**Vascular system** — The lumen of the dorsal vessel was lined by a thick, scalloped membrane without discernible lining endothelium which stains with aniline blue. Inner longitudinal and outer circular muscle layers about 14 μm thick were external to this membrane; the latter filled the scallops of the lining membrane (Figure 3E). The ventral vessel was slightly smaller than the dorsal and had a thinner wall with only longitudinal muscle; the lining was also thin and scalloped. The outer surface of the muscle was lined by mesothelium that rested on a thin basement membrane (Figure 3F). In experimental worms, the dorsal vessel was 12-μm thick; the longitudinal muscle was absent, while the circular was 5 to 8 μm and the lining membrane 2 to 3 μm, hyalinized, and highly scalloped (Figure 4E). The muscle of the ventral vessel was reduced but with greatly thickened membranes on both sides (Figure 4F).

**Excretory system** — The nephridia of degenerate worms were much more prominent because of increased numbers of interstitial cells; the majority were chloragogen cells, few were acidophil and fibroblast-like cells. These seemed to force the tubules apart. The epithelial linings of all the tubules were detached from the basement membranes and appeared as darkly staining masses. The basement membrane was thickened ten or more times by substances which stained with PAS and Congo red. Variable layers of circumferential fibroblast-like cells were found external to the hyalinized membranes; these had oval nuclei

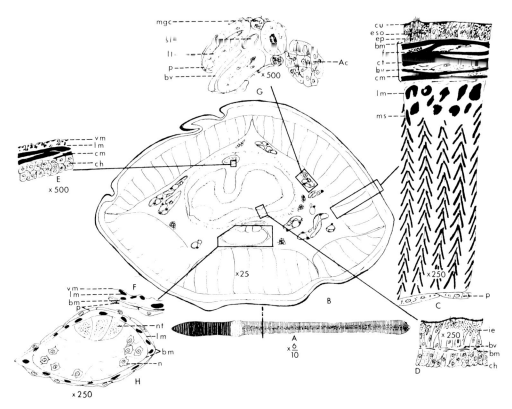

FIGURE 3.   Semidiagrammatic drawings of normal *Lumbricus*. 3A = gross, dorsal view × 6/10; B = schematic cross-sectional drawing of A behind the clitellum × 25; C = body wall addendum × 250; D = intestinal wall × 250; E = dorsal blood vessel × 500; F = ventral blood vessel × 250; G = nephridium × 500; and H = ventral nerve cord × 250.

## PLATE 1

| | | | | | |
|---|---|---|---|---|---|
| Ac | = | acidophil coelomic cells | lm | = | longitudinal muscle layer |
| bm | = | basement membrane | lt | = | large nephridial tuble |
| bv | = | small blood vessel | m | = | muscle fiber |
| ch | = | chloragogen cells or tissues | mgc | = | multinucleate giant cell |
| cm | = | circular muscle layer | ms | = | muscular septum |
| ct | = | connective tissue | n | = | neuron |
| cu | = | cuticular layer | nt | = | nerve trunk |
| ep | = | epidermis | p | = | peritoneum |
| eso | = | epidermal sense organ | | | |
| f | = | fibroblast or fibroblast-like | st | = | small nephridial tubule |
| ie | = | intestinal epithelium | vm | = | lining membrane of blood vessel |

and spindle-shaped bodies that were irregularly fibrillated. Chloragogen cells similar to those associated with the gut were found in the interstices of these structures. Moreover, polygonal cells with acidophilic, granular cytoplasm and the usual phagocytic, acidophilic, coelomic cells were present. Chloragogen or mesothelial cells lined the free surfaces of the nephridial masses. Blood vessels were not present, and multinucleate giant cells were strikingly absent compared to normals.[84] Nephridia in degenerate worms were found predominantly in the ventral region near the ventral nerve cord instead of in the lateral portion of the coelom (Figures 4B and G).

**Nervous system** — The normal long and short diameters of the ventral nerve cord were 260 and 100 μm, respectively. A thin muscular layer surrounded the cord and a connective

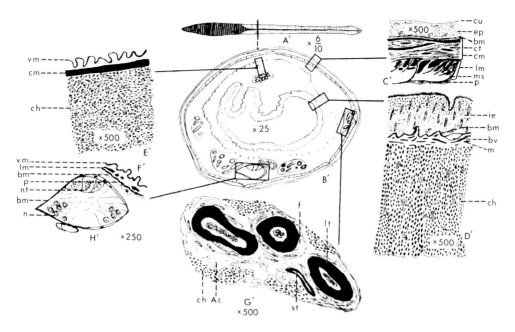

FIGURE 4. Semidiagrammatic drawings of aged *Lumbricus*. 4A′ = gross, dorsal view × 6/10; B′ = schematic cross sectional drawing of A′ behind the darker portion × 25; C′ = body wall × 500; D′ = intestinal wall × 500; E′ = dorsal blood vessel × 500; F′ = ventral blood vessel × 250; G′ = nephridium × 500; and H′ = ventral nerve cord × 250.

tissue membrane completely enveloped the muscle. The neurons were 10 μm in diameter with typical fish-eye-shaped nuclei and basophilic, stellate cytoplasm (Figure 3H). In the degenerate, the ventral nerve cord was 140 × 83 μm, the connective tissue membranes were thickened, and the neurons were only 5 μm in diameter with scanty cytoplasm (Figure 4H).

## C. What Factors Influence the Syndrome?
### 1. Natural Senescence vs. Biophysical Factors
What precisely causes runting and the histopathologic features of degeneration is not known. Our worms could have been close to actual "old age" or even much younger but prematurely caused to degenerate as a result of laboratory conditions. We did not know their actual ages when brought into the laboratory, but they were sexually mature and much older than 1 year, as judged by their sizes. Development from cocoon to adulthood requires 1 year,[85] and according to Comfort,[4,81] *Allolobophora* can live as long as 6 years, but worms operated on and kept under artificial conditions live shorter lives.[86] Light and moisture were controlled, albeit unlike those in nature, and in the wild, worms are free to move if the surrounding conditions become unlivable. We confined the worms in plastic boxes; however, they still seemed to feed since the gut was always full and fecal pellets present when the medium was changed. They foraged at night when the lights were off and remained quietly buried in the medium during the day when the lights were on. The relatively low ambient temperature seemed optimal; however, oxygen supply might have been inadequate.

### 2. Influence of the Laboratory Diet
The diet undoubtedly contributed to histopathologic changes so that the whole syndrome in earthworms may not be the picture of natural, slow aging as might be implied. This syndrome could be the result of malnutrition that triggered degeneration. Our medium, composed of Buss-Bedding, consists primarily of pulverized newspaper converted into mud-like consistency by mixing with water. The chemical composition of this is not known, but

carbohydrate (cellulose) is certainly one of its components. To obviate a suspected lack of nutritive elements, a mixture of walnut meal and peanut oil was added as a suggested supplement by our animal collector (Mr. G. J. Dougherty) who maintains earthworms temporarily in this manner prior to purchases by fishermen. Walnut meal is walnut kernel after about 90% of its lipid content is removed by processing at 180°C.[87] Protein content is 12 to 18%. It is important to note that the medium in which the worms were kept was concocted empirically; we do not know the actual nutritional requirements of earthworms, how much of the medium put in the boxes they ingest, and furthermore, whether or not they are metabolizing whatever they eat. However, this diet semed to have maintained them for only a short period after which they consistently lost weight. Highly proteinaceous diets apparently contribute to the formation of amyloid in vertebrates.[88] Whether or not the medium is highly proteinaceous with respect to the requirements of the worms is not known, but there was formation and deposition of amyloid, especially in the nephridia. This change in the nephridea could have led to inadequate disposal of metabolic waste products since these organs were responsible for excretion of one half of the nitrogenous wastes such as ammonia and urea.[89] Since chloragogen tissue may be involved in food storage and waste disposal like a primitive liver, the histopathologic changes seen in degenerate worms might be compensatory. If the underlying cause of this degenerative process were malnutrition, then the picture was that of "fatty metamorphosis."

The walnut meal contains tocopherols. It is unlikely to be present in Buss-Bedding while peanut oil has 11.2 mg %.[90] With the amount of peanut oil added to the medium, Vitamin E is practically nil, and its deficiency could be one of the causes of the muscular atrophy.[91] Thus, inefficient disposal might lead to accumulation of wastes, resulting in eventual systemic poisoning. Moreover, normal excretion may have been severely hampered by derangement of the integument, another site for elimination in annelids. Although the etiology of this syndrome is unknown, it still provides a feasible model for experimentally analyzing degeneration, which superficially mimics some aging changes in longer-living higher vertebrates.

## VII. NEW APPROACHES TO WASTE MANAGEMENT AND LIFE SPANS AMONG CERTAIN OLIGOCHAETE ANNELIDS (LUMBRICIDS)

Turning to the more practical aspects of worm viability, newer studies on earthworms have focused on other aspects of worm lifespan. Castings produced from a mineral soil by *Eisenia foetida* were toxic to itself, to *Eudrilus eugeniae*, and two *Amyntha spp.* In contrast, castings produced from the soil by these two latter genera allowed a gain or maintenance of weight for all three earthworms.[92] Other work has dealt with sludge but with respect to its decomposition by *Eisenia*.[93] These results may have implications for lifespan and waste management using earthworms.

## VIII. STUDIES ON AGING, POPULATION DENSITY, REPRODUCTION, AND SIZE

Small founder-units of 2 to 4 worms each of adult *Allolobophora caliginosa* (Sav.) and *Lumbricus rubellus* (Hoffm.) still contained live adults after 14 months (82% of the *A. caliginosa* and 35% of *L. rubellus* units). *A. caliginosa* also produced significantly more cocoons and newly hatched individuals than did *L. rubellus,* a success attributed to relatively long adult lifespan rather than having a high birth rate.[94] In *A. rosea,* the construction of composite growth curves indicated that most worms attain sexual maturity at age 2 years, and that field specimens of 330 to 350 mg fresh-weight were probably 5 to 6 years old.[95] The estimated age of one particularly large specimen (439.4 mg fresh-weight) was probably 8 years. In field cultures, 85% of the annual cocoon production occurred during spring and summer. Age at which reproduction terminates and hatchability have been studied in *Eisenia foetida* with respect to optimum population density. A family of regression equations was formulated taking into consideration time of clitellum formation and number of cocoons produced per adult in relation to age and population density.[96] Obviously, data generated from these approaches will be useful in designing breeding studies in relation to age.

## IX. POSSIBLE ROLES OF COELOMIC CELLS IN ANNELID REGENERATION AND AGING

Much of this chapter has focused on the coelomic cells, their structure, and function in relation to the earthworm defense and immune system. These studies have not yet shown clear relationship between aging and the immune system[97] (except for chloragogue cells). Cloragogue cells, according to Roots,[98] are comparable to the liver due to the presence of glycogen, phospholipids, and urea. Ferretin, hemoglobin, and flavins have also been observed. At least in *Eisenia*, these cells are known to be the source of a lytic substance EFAF.[99] They too are a component of the worm's defense system. Chloragogue cells are bright yellow in young worms and in many instances, a color gradient can be observed through the body wall of *Stylaria*. The globules of such cells will appear, however, deep yellow, orange, and brown. In the oldest, darkest regions, the enigmatic pigment of aging tissues, lipofuscin, has been observed.[100] This raises the question of whether there is a precise correlation between aging in worms, as caused by chloragogue cells, or whether these cells alone undergo age changes.

## X. FINAL COMMENT

Biomedical research is greatly aided by the use of relevant animals as models for analyzing the mechanisms of disease in humans. While metazoans undergo aging and possess an immune system, studies linking immunity and aging have not been systematically performed using such species.[101] Regardess of lifespan, many of these animals can be made to live still longer by maintenance at lower temperature. Although research has been limited, it has been shown that controlled undernutrition substantially prolongs lifespan in nematode worms.[102] Of economic importance is the common earthworm with a defined immune system, an extended lifespan in some species, and degenerative senescent-type changes during prolonged laboratory maintenance. This chapter has raised several questions related to understanding lifespan using earthworms as models.

## REFERENCES

1. **Meglitsch, P. A.,** *Invertebrate Zoology,* Oxford University Press, N.Y., 1967, 961.
2. **Gabriel, M. L. and Fogel, S.,** *Great Experiments in Biology,* Prentice-Hall, Englewood Cliffs, N.J., 1955, 317.
3. **Cooper, E. L.,** *Comparative Immunology,* Prentice-Hall, Englewood Cliffs, N.J., 1976, 338.
4. **Comfort, A.,** *The Process of Aging,* New American Library of World Literature, N.Y., 1961, 144.
5. **Strehler, B. L.,** *Time Cells and Aging,* Academic Press, N.Y., 1962, 270.
6. **Walford, R. L.,** *The Immunologic Theory of Aging,* Munksgaard, Copenhagen, 1969, 248.
7. **Hildemann, W. H., Clark, E. A., and Raison, R. L.,** *Comprehensive Immunogenetics,* Elsevier/North-Holland, N.Y., 1981, 368.

8. **Cooper, E. L.,** Immunity in Invertebrates, *CRC Rev. Immunol.,* 29, 1, 1981.

9. **Cooper, E. L.,** Evolution of blood cells, *Ann. Immunol. (Inst. Pasteur),* 127c, 817, 1976.

10. **Cooper, E. L. and Stein, E. A.,** Oligochaetes, in *Invertebrate Blood Cells,* Ratcliffe, N. A. and Rowley, A. F., Eds., Academic Press, New York, 1981, *15.*

11. **Stein, E. A. and Cooper, E. L.,** Agglutinins as receptor molecules: a phylogenetic approach, in *Developmental Immunology, Clinical Problems and Ageing,* Cooper, E. L. and Brazier, M. A. B., Eds., Academic Press, N.Y., 1982, 85.

12. **Child, C. M.,** *Senescence and Rejuvenescence,* University of Chicago Press, Chicago, 1915.

13. **Orton, J. H.,** Reproduction and death in invertebrates and fishes, *Nature (London),* 123, 14, 1929.

14. **Szabo, I.,** Senescence and death in invertebrate animals, *Riv. Biol. Firenze,* 19, 377, 1935.

15. **Cameron, G. R.,** Inflammation in earthworms, *J. Pathol. Bacteriol.,* 35, 933, 1932.

16. **Barnes, R. D.,** *Invertebrate Zoology,* W. B. Saunders, Philadelphia, 1968, 743.

17. **Valembois, P.,** Role des leucocytes dans l'acquisition d'une immunite antigreffe specifique chez les lombriciens, *Arch. Zool. Exp. Gen.,* 112, 97, 1971a.

18. **Valembois, P.,** Etude ultrastructurale des coelomocytes du lombricien *Eisenia foetida, Bull. Soc. Zool. Fr.,* 96, 59, 1971b.

19. **Valembois, P.,** Cellular aspects of graft rejection in earthworms and some other metazoa, in *Contemporary Topics in Immunobiology,* Cooper, E. L., Ed., Plenum Press, N.Y., 1974.

20. **Hostetter, R. K. and Cooper, E. L.,** Earthworm coelomocyte immunity, in *Contemporary Topics in Immunobiology,* Cooper, E. L., Ed., Plenum Press, N.Y., 1974.

21. **Lemmi, C. A., Cooper, E. L., and Moore, T. C.,** An approach to studying evolution of cellular immunity, in *Contemporary Topics in Immunobiology,* Cooper, E. L., Ed., Plenum Press, N.Y., 1974.

22. **Cooper, E. L.,** The earthworm coelomocyte. A mediator of cellular immunity, in *Phylogeny of Thymus and Bone Marrow - Bursa Cells,* Wright, R. K. and Cooper, E. L., Eds., Elsevier/North-Holland, N.Y., 1976b.

23. **Cooper, E. L.,** Cellular recognition of allografts and xenografts in invertebrates, in *Comparative Immunology,* Marchalonis, J. J., Ed., Blackwell Scientific, Oxford, 1976c.

24. **Kindred, J. E.,** The leucocytes and leucopoietic organs of an oligochaete, *Pheretima indica* (Horst), *J. Morphol. Physiol.,* 47, 435, 1929.

25. **Liebmann, E.,** The coelomocytes of Lumbricidae, *J. Morphol.,* 71, 221, 1942.

26. **Duprat, P. and Bouc-Lassalle, A. M.,** Mise au point et etude du liquide coelomique du Lombriciens *Eisenia foetida, Bull. Soc. Zool. R.,* 92, 767, 1967.

27. **Darwin, C.,** *The Formation of Vegetable Mould Through The Action of Worms With Observations on Their Habits,* D. Appleton-Century-Crofts, N.Y., 1900, 326.

28. **Manning, M. J. and Turner, R. J.,** *Comparative Immunobiology,* Halstead Wiley, 1976, 184.

29. **Marchalonis, J. J., Ed.,** *Comparative Immunology,* Blackwell Scientific, Oxford, 1976, 470.

30. **Gershwin, M. E. and Cooper, E. L., Eds.,** *Animal Models of Comparative and Developmental Aspects of Immunity and Disease,* Pergamon Press, N.Y., 1978, 396.

31. **Cooper, E. L., Ed.,** *Contemporary Topics in Immunobiology,* Vol. 4, Plenum Press, N.Y., 1974, 299.

32. **Hildemann, W. H. and Benedict, A. A., Eds.,** *Advances in Experimental Medicine and Biology,* Vol. 64, Plenum Press, N.Y., 1975, 485.

33. **Wright, R. K. and Cooper, E. L.,** *Phylogeny of Thymus and Bone-Marrow-Bursa Cells,* Elsevier/North-Holland, Amsterdam, 1976, 325.

34. **Cooper, E. L.,** *Developmental Immunobiology,* Solomon, J. B. and Horton, J. D., Eds., Elsevier/North-Holland, Amsterdam, 1977, 456.

35. **Cooper, E. L.,** What is the origin and function of the immune system?, in *Trends in Biochemical Sciences,* 2, 130, 1977.

36. **Garland, J. M.,** The T-cell paradigm. A philosophical view of immunology, *Dev. Comp. Immunol.,* 2, 39, 1978.

37. **Warr, G. W.,** On T-cells in invertebrates, *Dev. Comp. Immunol.,* 2, 555, 1978.

38. **Cooper, E. L. and Hildemann, W. H.,** *Phylogeny Workshop, Prog. Immunol.,* 3, 1977, 738.

39. **Cooper, E. L., Ed.,** *Contemporary Topics in Immunobiology,* Vol. 4, Plenum Press, N.Y., 1974, 299.

40. **Stein, E. A., Wojdani, A., and Cooper, E. L.,** Naturally-occurring and induced hemagglutinins in the coelomic fluid of the earthworm, *Lumbricus terrestris, Am. Zool.,* 20, 864, 1980.

41. **Stein, E. A., Wojdani, A., and Cooper, E. L.,** Partial characterization of agglutinins from the earthworm, *Lumbricus terrestris, Am. Zool.,* 21, 974, 1981.

42. **Stein, E. A., Wojdani, A., and Cooper, E. L.,** Agglutinins in the earthworm, *Lumbricus terrestris:* naturally occurring and induced, *Dev. Comp. Immunol.,* in press.

43. **Wojdani, A., Stein, E. A., Lemmi, C. A., and Cooper, E. L.,** Agglutinins and proteins in the earthworm, *Lumbricus terrestris,* before and after injection of erythrocytes, carbohydrates and other materials, *Dev. Comp. Immunol.,* 6, 613, 1982.

44. **Cooper, E. L.,** *Comparative Pathobiology,* Vol. 3, Bulla, L. A., Jr. and Cheng, T. C., Eds., Plenum Press, N.Y., 1977, 192.
45. **Duprat, P.,** Evidence de reactions immunitaires dans les homogreffes de paroi du corps chez le lombricien *Eisenia foetida typica, C. R. Acad. Sci., Ser. D.,* 259, 4177, 1964.
46. **Duprat, P.,** Étude de la prise et du maintien d'un greffon de paroi du corps chez le lombricien *Eisenia foetida, Ann. Inst. Pasteur,* 113, 867, 1967.
47. **Cooper, E. L.,** Transplantation immunity in annelids. I. Rejection of xenografts exchanged between *Lumbricus terrestris* and *Eisenia foetida, Transplantation,* 6, 322, 1968.
48. **Cooper, E. L.,** Chronic allograft rejection in *Lumbricus terrestris, J. Exp. Zool.,* 171, 69, 1969.
49. **Cooper, E. L.,** Specific tissue graft rejection in earthworms, *Science,* 166, 1414, 1969b.
50. **Valembois, P.,** Étude d'une Hétérogreffe de Paroi du Corps chez les Lombriciens, *Ph.D. thesis,* University of Bordeaux, Talence, France, 1970.
51. **Cooper, E. L.,** Rejection of body wall xenografts exchanged between *Lumbricus terrestris and Eisenia foetida, Am. Zool.,* 5, 169, 1965.
52. **Cooper, E. L.,** A method of tissue grafting in the earthworm, *Lumbricus terrestris, Am. Zool.,* 5, 233, 1965.
53. **Cooper, E. L. and Rubilotta, L.,** Allograft rejection in *Eisenia foetida, Transplantation,* 8, 220, 1969.
54. **Valembois, P.,** Étude anatomique de l'évolution de greffons hétéroplastiques de paroi du corps chez quelques lombriciens, *C. R. Acad. Sci., Ser. D.,* 257, 3227, 1963.
55. **Duprat, P. C.,** Specific allograft reactions in *Eisenia foetida, Transplant. Proc.,* 3, 222, 1970.
56. **Hostetter, R. and Cooper, E. L.,** Coelomocytes as effector cells in earthworm immunity, *Immunol. Commun.,* 1, 155, 1972.
57. **Parry, M. J.,** Survival of body wall autografts, allografts and xenografts in the earthworm *Eisenia foetida, J. Invertebr. Pathol.,* 31, 383, 1978.
58. **Linthicum, D. S., Marks, D. H., Stein, E. A., and Cooper, E. L.,** Graft rejection in earthworms: an electron microscopic study, *Eur. J. Immunol.,* 7, 871, 1977b.
59. **Marks, D. H., Stein, E. A., Cooper, E. L., and Ramirez, J. A.,** Accelerated graft rejection in earthworms: immunologic memory?, *J. Invertebr. Pathol.,* submitted.
60. **Cooper, E. L. and Roch, P.,** Second-set allograft responses in the earthworm *Lumbricus terrestris:* kinetics and characteristics, *Transplantation,* submitted.
61. **Valembois, P.,** Liberation de phosphatase acide dans les cellules musculaires d'un greffon de paroi du corps chez un lombricien, *J. Microsc.,* 7, 61a, 1968.
62. **Valembois, P.,** Cellular aspects of graft rejection in earthworms and some other metazoa, in *Contemporary Topics in Immunobiology,* Vol. 4, Cooper, E. L., Ed., Plenum Press, N.Y., 121, 1974.
63. **Hostetter, R. K. and Cooper, E. L.,** Cellular anamnesis in earthworms, *Cell. Immunol.,* 9, 384, 1973.
64. **Bailey, S., Miller, B. J., and Cooper, E. L.,** Transplantation immunity in annelids. II. Adoptive transfer of the xenograft reaction, *Immunology,* 21, 81, 1971.
65. **Valembois, P.,** Rôle des leucocytes dans l'acquisition d'une immunité antigreffe spécifique chez les lombriciens, *Arch. Zool. Exp. Gen.,* 112, 97, 1971a.
66. **Lemmi, C. A. E. and Cooper, E. L.,** Induction of coelomocyte proliferation by xenografts in the earthworm *Lumbricus terrestris, Dev. Comp. Immunol.,* 5, 73, 1981.
67. **Roch, P., Valembois, P., and DuPasquier, L.,** Response of earthworm leukocytes to concanavalin A and transplantation antigens, *Adv. Exp. Med. Biol.,* 64, 45, 1975.
68. **Valembois, P. and Roch, P.,** Identification par autoradiographie des leucocytes stimulés à la suite de plaies ou de greffes chez un ver de terre, *Biol. Cell.,* 28, 81, 1977.
69. **Marks, D. H., Stein, E. A., and Cooper, E. L.,** Chemotactic attraction of *Lumbricus terrestris* coelomocytes to foreign tissue, *Dev. Comp. Immunol.,* 3, 277, 1979.
70. **Klein, J.,** Evolution and function of the major histocompatibility system: facts and speculations, in *The Major Histocompatibility System in Man and Animals,* Götze, D., Ed., Springer-Verlag, Berlin, 1977, 340.
71. **Cohen, N. and Collins, N. H.,** Major and minor histocompatibility systems of ectothermic vertebrates, in *The Major Histocompatibility System in Man and Animals,* Götze, D., Ed., Springer-Verlag, Berlin, 1977, 313.
72. **Dupasquier, L.,** Phylogenesis of vertebrate immune system, in *The Immune System,* Melchers, F. and Rajewsky, R., Eds., Springer-Verlag, Berlin, 1976, 101.
73. **Schreffler, D. C., David, C. S., Passmore, H. C., and Klein, J.,** Genetic organization and evolution of the mouse H-2 region: a duplication model, *Transplant. Proc.,* 3, 1971, 176.
74. **Cooper, E. L.,** Phylogenetic aspects of transplantation, in *Handbuch der Allgemeinen Pathologie, VI/8, Transplantation,* Masshoff, J. W., Ed., Springer-Verlag, Berlin, 1977, 139.
75. **Ohno, S.,** *Evolution by Gene Duplication,* Springer-Verlag, Berlin, 1970.
76. **Berggard, I. and Bearn, A. G.,** Isolation and properties of a low molecular weight β-globulin occurring in human biological fluids, *J. Biol. Chem.,* 243, 4095, 1968.

77. Nakamura, K., Tanigaki, N., and Pressman, D., Multiple common properties of human $\beta_2$-microglobulin and the common portion fragment derived from HL-A antigen molecules, *Proc. Natl. Acad. Sci. U.S.A.*, 70, 2863, 1973.

78. **Peterson, P. A., Cunningham, B. A., Berggard, I., and Edelman, G. M.,** $\beta_2$ Microglobulin — a free immunoglobulin domain, *Proc. Natl. Acad. Sci. U.S.A.*, 69, 1697, 1972.

79. **Shalev, A., Greenberg, A. H., Lögdberg, L., and Bjorck, L.,** $\beta_2$-Microglobulin-like molecules in low vertebrates and invertebrates, *J. Immunol.*, 127, 1186, 1981.

80. **Roch, P. and Cooper, E. L.,** A $\beta_2$-microglobulin-like molecule on earthworm (*L. terrestris*) leukocyte membranes, *Dev. Comp. Immunol.*, 7, 633, 1983.

81. **Comfort, A.,** *The Biology of Senescence*, Rinehart & Co., N.Y., 1956, 257.

82. **Hylton, A. R.,** Histopathological changes with age in the flight muscle of the African mosquito *Eretmapodites chrysogaster*, *J. Invertebr. Pathol.*, 8, 75, 1966.

83. **Cooper, E. L. and Baculi, B. S.,** Degenerative changes in the annelid *Lumbricus terrestris*, *J. Gerontol.*, 23, 375, 1968.

84. **Cooper, E. L.,** Multinucleate giant cells, granulomata and myoblastomas in annelid worms, *J. Invertebr. Pathol.*, 11, 123, 1968.

85. **Stephenson, J.,** *The Oligochaeta*, Clarendon Press, Oxford, 1930, 978.

86. **Rabes, O.,** Ueber Transplantationsversuche an lumbriciden, *Biol. Zbl.*, 21, 633, 1901.

87. **Johnson, R. A.,** *Physical Properties and Chemical Analyses of Walnut Kernels and By-Products*, Diamond Walnut Growers, Stockton, California, 1968.

88. **Robbins, R. L.,** *Textbook of Pathology*, W. B. Saunders, Philadelphia, 1967, 1190.

89. **Laverack, M. S.,** *The Physiology of Earthworms*, Pergamon Press, N.Y., 1963, 206.

90. **Jacobs, M. B., Ed.,** *The Chemistry and Technology of Food and Food Products*, Vol. 2, Wiley-Interscience, N.Y., 1951.

91. **Berneske, G. M., Butson, A. R. C., Gauld, E. N., and Levy, D.,** Clinical trial of high dosage vitamin E in human muscular dystrophy, *Can. Med. Assoc. J.*, 82, 418, 1960.

92. **Kaplan, D. L., Hartenstein, R., and Neuhauser, E. F.,** Coprophagic relations among the earthworms *Eisenia foetida, Eudrilus eugeniae* and *Amynthas*- SPP, *Pedobiologia*, 20, 74, 1980.

93. **Mitchell, M. J.,** Functional relationships of macro-invertebrates in heterotrophic systems with emphasis on sewage sludge decomposition, *Ecology*, 60, 1270, 1979.

94. **Bengtson, S -A., Nilsson, A., Nordstrom, S., and Rundgren, S.,** Short-term colonization success of lumbricid founder populations, *Oikos*, 33, 308, 1979.

95. **Phillipson, J. and Bolton, P. J.,** Growth and cocoon production by *Allolobophora rosea*, Oligochaeta, Lumbricidae, *Pedobiologia*, 17, 70, 1977.

96. **Hartenstein, R. Neuhauser, E. F., and Kaplan, D. L.,** Reproductive potential of the earthworm *Eisenia foetida*, *Oecologia*, 43, 329, 1979.

97. **Moment, G. B.,** The possible roles of coelomic cells and their yellow pigment in annelid regeneration and aging, *Growth*, 38, 209, 1974.

98. **Roots, B. I.,** Some observations on the chloragogue tissue of earthworms, *Comp. Biochem. Physiol.*, 1, 218, 1960.

99. **Roch, P., Valembois, P., Divant, N., and Lessegues, M.,** Protein analysis of earthworm coelomic fluid. II. Isolation and biochemical characterization of the *Eisenia foetida* andrei Factor (EFAF), *Comp. Biochem. Physiol.*, 69b, 829, 1981.

100. **Hendley, D. D. and Strehler, B. I.,** Enzymatic activities of lipofuscin age pigments: comparative histochemical and biochemical studies, *Biochem. Biophys. Acta*, 99, 406, 1965.

101. **Cooper, E. L. and Walford, R. L.,** New perspectives on aging and immunity: lower animals, ontogeny, phylogeny and immunoendocrinology, *Dev. Comp. Immunol.*, 6, 391, 1982.

102. **Klass, M. R.,** Aging in the nematode *Caenorhabditis elegans:* major biological and environmental factors influencing life span, *Mech. Aging Dev.*, 6, 413, 1977.

Chapter 7

# THE USE OF SHELL MICROGROWTH PATTERNS IN AGE DETERMINATIONS OF THE HARD CLAM, *MERCENARIA MERCENARIA* (LINNÉ)

## Michael J. Kennish

## TABLE OF CONTENTS

# I. INTRODUCTION

The hard clam, *Mercenaria mercenaria* Linné, is an infaunal bivalve that inhabits coastal embayments in North America from the Gulf of St. Lawrence to the Gulf of Mexico. It has been successfully introduced into France,[1] the British Isles,[2,3] and Holland.[4] The organism burrows in sediments of the intertidal and subtidal zones, and is confined to waters less than 10 m in depth where salinities range from 15 to 35‰.[5,6]

*M. mercenaria* supports one of the most valuable commercial and recreational mollusc fisheries in the U.S., with maximum yields occurring from Cape Cod to Cape Hatteras. Much of the annual harvest of the hard clam is from New York waters.[7] In 1977, commercial production in the eastern U.S. totaled about 6045 metric tons of meats, having a dollar value in excess of $25 million.[8] Fishermen harvest the hard clam almost exclusively from wild populations using a variety of equipment, but principally hand-held tongs and rakes. When individuals reach 3 to 4 cm in length at an age range of about 2 to 5 years, they enter the commercial and recreational fisheries.

Hard clams can be cultured in the laboratory. Methods of conditioning and rearing larvae and juveniles have been reviewed by Loosanoff and Davis.[9] Since the publication of that work, larvae and juveniles have been routinely reared for laboratory and field studies, and hatcheries have become established.

Because of its economic importance, *M. mercenaria* has been the subject of numerous biological investigations. Much research has centered around the early ontogenetic stages of the mollusc when it is most vulnerable to the vagaries of environmental conditions and subject to substantial predation resulting in high mortality rates.[10,11] Recently, the hard-clam shell has been scrutinized in ecological and paleoecological studies.[12-27] It has been shown to effectively record changes in the marine environment and the organism's physiology throughout ontogeny. Shell microgrowth patterns, for example, reflect the effects of temperature, tides, sediment, water depth, salinity, storms, food supply, spawning and age.[20] Thus, examination of the shell of this species has revealed valuable information for reconstructing habitat conditions in both extant and ancient marine environments. The shell has also yielded data on the population dynamics of the bivalve which cannot be obtained by other means.

The objective of this chapter is to demonstrate the usefulness of the shell of *M. mercenaria* in absolute age determinations. In order to attain this objective, the ecology and life history of the organism will be discussed initially since characteristics of the shell, such as microgrowth patterns and structure, are dependent on conditions in the ambient environment as well as the bivalve's physiology. Subsequently, I will give a detailed description of the shell itself, and show how shell growth patterns can serve to determine age of individual specimens from natural populations.

# II. ECOLOGY AND LIFE HISTORY

*M. mercenaria* attains sexual maturity at an age of 2 years in most regions;[14] it becomes sexually mature in 1 year in Florida waters.[28] The species is protandrous. During the first year of life, all members of a population contain male gonads, but approximately half of them undergo sex reversal by the end of their 2nd year of life.[29,30] The hard clam spawns from spring to fall throughout its geographical range when water temperature surpasses 20°C.[31,32] Spawning is a group reaction stimulated in part by water temperature[10] and by a pheromone carried in the sperm.[33] Davis and Chanley[34] noted that the total number of eggs released per season by a female varies from 8 to 39.5 million, with an average of 24.6 million. Bricelj and Malouf,[35] however, observed lower fecundity values, with the maximum egg production by a single female per spawning period being $16.8 \times 10^6$ eggs. Larger clams (cherrystones, 36.5 to 41.3 mm in shell width; chowders, greater than 41.3 mm) produced significantly more eggs than smaller clams (littlenecks, 25.4 to 36.5 mm in shell width).

Fertilization occurs external of the the female subsequent to the release of gametes into the environment. After fertilization, an individual passes through a series of meroplanktonic (larval) stages prior to metamorphosis into an adult form. These stages are characterized by distinct anatomical development of the velum, byssus, siphons, and valves, as well as changes in the position of the animal in the habitat.[10] Table 1 lists these ontogenetic stages; Chanley and Andrews[36] provide photographic sequences of them. Early in development, a larva floats passively in the water column, but with the growth of the velum it becomes a strong swimmer. Hard-clam larvae have been observed to swim upward at a rate of 8 cm/min.[37]

Larvae of *M. mercenaria* are planktotrophic, consuming large amounts of phytoplankton, particularly naked diatoms and dinoflagellates. Davis and Guillard[38] reported that *Monochrysis lutheri* and *Isochrysis galbana* constituted an excellent food source for larvae of the hard clam, and Carriker[10] found rapid growth of larvae which were fed a mixed culture of either *Nannochloris* or *Chlorella*. The food requirements of the larvae of this species appear to be less restricted than other bivalves. Loosanoff and Davis[9] identified more than ten algal species utilized as food by larvae of *M. mercenaria;* however, of these ten species, *Monochrysis lutheri, Isochrysis galbana, Dicrateria* sp., *Chlorococcum* sp., and *Platymonas* sp. had the greatest food value.

Eggs and larvae of *M. mercenaria* are subject to a large wastage of numbers due to disease, predation, and extremes in environmental conditions. Many planktonic, pelagic, and benthic organisms feed on meroplankton such as the larvae of *M. mercenaria*.[39] Bacterial and fungal infections can retard growth and increase mortality. Under laboratory conditions, when subjected to water temperatures below 15°C and above 33°C, fertilized eggs and larvae show abnormal development and heavy mortality, with early cleavage stages having the narrowest range of temperature tolerance.[9,40] In Little Egg Harbor, New Jersey, Carriker[10] estimated larval survival to be 2.6%.

The duration of the larval period varies considerably. Individuals mature and metamorphose at different rates depending on environmental factors (e.g., food supply and water temperature) and, in part, on inherited characteristics.[41] The setting of larvae on a substratum generally occurs from 1 week to 1 month after fertilization. It is enhanced by higher water temperatures. Loosanoff et al.[42] revealed that at 30°C, setting began 7 days and ended 16 days after fertilization, but at 18°C setting began 16 days and ended 24 days after fertilization.

As *M. mercenaria* undergoes metamorphosis, a shell, siphon, and foot develop, and the bivalve burrows into sediment composed most commonly of sand, silt, and clay. Metamorphosis may be delayed until a favorable habitat is found.[43] Larvae of *M. mercenaria* prefer to set on silt and sand bottoms rather than clay, gravel, and shell bottoms.[10,44] Growth is most rapid in sand.[45,46]

During their first year of life in sediment, many hard clams are lost to predators, principally crabs, snails, and fishes. Important predators include the blue crab (*Callinectes sapidus*,) green crab ( *Carcinus maenas*,) horseshoe crab (*Limulus polyphemus*,) rock crab (*Cancer irroratus*,) mud crab (*Neopanopeus sayi*,) boring gastropods (*Polinices duplicatus, Lunatia heros, Urosalpinx cinerea*, and *Eupleura caudata*,) whelks (*Busycon carica* and *B. canaliculatum*,) sea star (*Asterias forbesi*,) rays (*Dasyatidae* and *Myliobatidae*,) summer flounder (*Paralichthys dentatus*,) tautog (*Tautoga onitis*,) and northern puffer (*Sphoeroides maculatus*.)[8,10,11,47] Predation can quickly decimate an entire population of newly set clams. MacKenzie[11] showed that natural populations of *M. mercenaria* are most numerous where predators are scarce or stones provide coverage for them. Setting density and predation determine relative clam abundance. Predation pressure decreases as hard clams grow and their shells thicken. When individuals exceed 50 mm in length, their major predator becomes man.

**Table 1**
### EARLY ONTOGENETIC STAGES OF THE HARD CLAM, *MERCENARIA MERCENARIA*

| Stage | Description | Age |
|---|---|---|
| Nonshell planktonic | Fertilized ovum, blastula, gastrula, tro-chophore, nonshell veliger | 0—48 hr. |
| Straight-hinge veliger | Smooth valves with velum, length 90—140 μm | 1—3 days |
| Umbone veliger | Umbone projects above hinge line, length 140—220 μm | 3—20 days |
| Pediveliger | Organism possesses a velum and foot and alternates swimming in water and crawling on sediment, length 170—230 μm | 6—20 days |
| Byssal plantigrade | Bivalve affixes to the substratum by means of a byssus, it alternates crawling and byssal attachment, siphons form, length 230 μm—9mm | >20 days |
| Juvenile plantigrade | Byssus lost, animal burrows into sediment by means of a foot, siphons well-developed, length >9 mm | |

Modified from Carriker, M. R., *J. Elisha Mitchell Sci. Soc.*, 77, 168, 1961.

The rate of growth of *M. mercenaria* is strongly dependent on water temperature and food supply. Because these factors vary within embayments and throughout the geographical range of the species, growth rates of hard clams differ significantly both within and among populations. Davis and Calabrese[48] discerned maximum larval growth at temperatures ranging from 25 to 30°C and salinities ranging from 20 to 27‰ Adults grow most rapidly at temperatures of 20 to 23°C and salinities of 20 to 23‰[4,49] According to Ansell,[4] the lower- and upper-temperature thresholds of stress on growth of *M. mercenaria* are 9 and 31°C, respectively. Although the thermal death-point for this species has been placed at 45.2°C,[50] individuals purportedly survive water temperatures as high as 34°C for only short periods of time.[51] Populations continuously exposed to water temperatures above 31°C probably experience increased mortality rates.

Growth is fastest during the first 3 or 4 years of life and thereafter declines gradually with increasing age. Kennish and Loveland[24] have described the nonlinear growth pattern of *M. mercenaria* in Barnegat Bay, New Jersey (Figure 1). Ontogenetic growth of these clams conforms best to the Gompertz growth equation:

$$H_t = 74.22_e^{-1.79_e^{-0.424}} \qquad (1)$$

where $H_t$ is the size at some time (t), 74.22 is the maximum or asymptotic size, (e) is the base of the natural logarithm, $-1.79$ is the value defining the proportion of $H_{max}$ which specifies the initial condition $H_O$ and $-0.424$ is the intrinsic growth factor.

Most hard clams live less than 10 years and grow to less than 80 mm in shell height and length. However, a maximum longevity of 25 to 40 years of age is attainable.[52,53] The pattern of ontogenetic mortality in the bivalve is high-low-high.[22] Mortality is high during the planktonic larval and pediveliger stages, lower subsequent to the plantigrade stages (Table 1), and high in the gerontic stage (Figure 2).

## III. SHELL GROWTH AND FORM

The shell of *M. mercenaria* consists of two symmetrical, ovate or heart-shaped valves attached at the dorsal margin by a hinge ligament. Figure 3 illustrates the major morphological

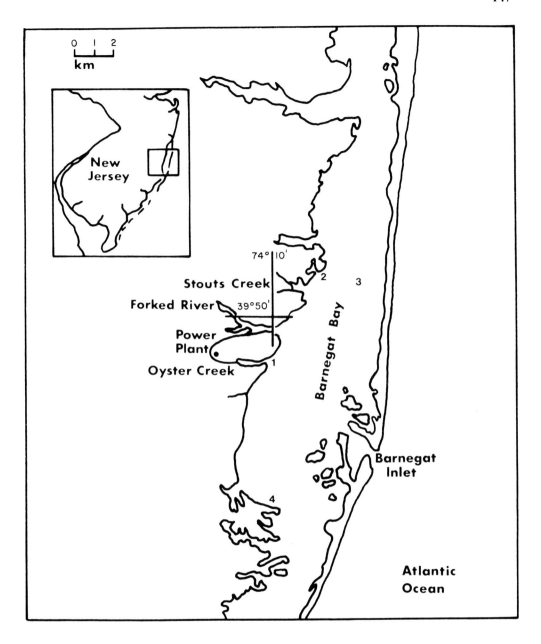

FIGURE 1.   Map of Barnegat Bay, N.J., showing sampling sites (1 to 4) of *Mercenaria mercenaria*. The inset at the upper left shows the location of Barnegat Bay in relationship to the state of New Jersey. (From Kennish, M. J., *Environ. Geol.*, 2, 223, 1978. With permission.)

features of the shell. A raised, convex region in each valve, located on either side of the elastic ligament, is termed the umbo (plural: umbos, umbones). The umbo terminates in a point, the beak, which is directed anteriorly. A heart-shaped lunule occupies the area immediately anterior and ventral to the beak, and an elongate escutcheon lies posterior to it. Concentric growth lines extend across the exterior of the valves, being most conspicuous near the margins rather than the central portions.

The interior of the shell, containing various markings, provides information on the structure of the soft parts of the animal. A pallial line parallel to the shell margin signifies the point of attachment of the shell-secreting mantle lobes. Along the posterior margin, the pallial

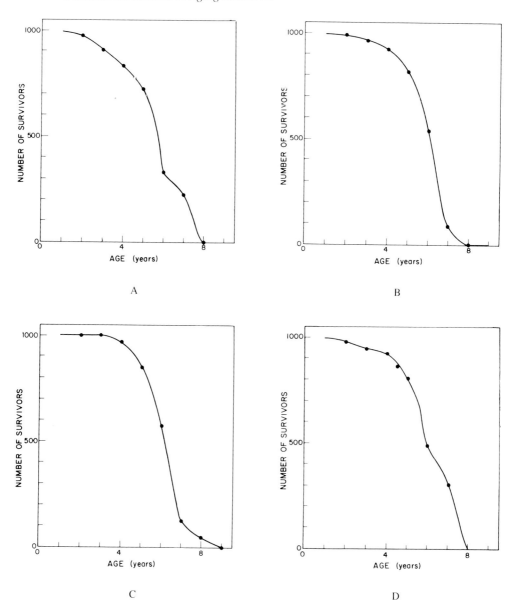

FIGURE 2.    (A to D) Survivorship curves for natural populations of *Mercenaria mercenaria* at sites 1 to 4 in Barnegat Bay. (A) = site 1; (B) = site 2; (C) = site 3; (D) = site 4. Data are plotted as the number of clams surviving at the beginning of each age internal ($l_x$) from an initial cohort of 1000 specimens. (From Kennish, M. J., *Environ. Geol.*, 2, 223, 1978. With permission.)

line forms a U-shaped pallial sinus, indicating the presence of siphons and siphonal retractor muscles. Anterior and posterior muscle scars appear as round impressions on the interior shell surface and are produced by the adductor muscles. When contracted, the adductor muscles oppose the ligament and close the valves. During this process, hinge teeth maintain proper alignment of the valves and preclude rotational and shearing movements.[54] The hinge teeth occur along the hinge which is situated below the umbos. They are differentiated into prominant cardinals, immediately below the umbos, and less obvious laterals, anterior and posterior to the cardinals. The valves of *M. mercenaria* open by the action of the ligament when the adductor muscles relax or the organism dies.

**a**

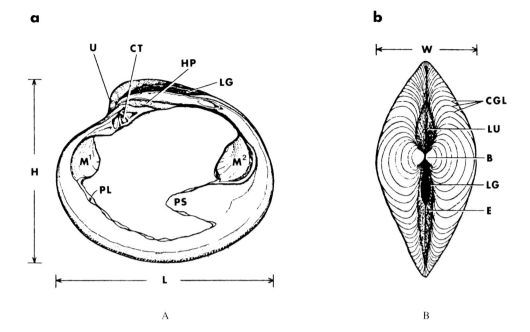

A

B

FIGURE 3.   Shell form and structural features of the hard clam, *Mercenaria mercenaria*. (A) L = length; H = height; U = umbo; CT = cardinal teeth; HP = hinge plate; LG = ligament; PL = pallial line; PS = pallial sinus; $M^1$ = anterior adductor muscle scar; $M^2$ = posterior adductor muscle scar. (B) W = width; CGL = concentric growth lines; LU = lunule; B = beak; LG = ligament; E = escutcheon. (Modified from Seed, R., in *Skeletal Growth of Aquatic Organisms*, Rhoads, D. C. and Lutz, R. A., Eds., Plenum Press, N.Y., 1980, 23. With permission.)

The shell is composed primarily of calcium carbonate (aragonite) and conchiolin (organic material). When the valves of *M. mercenaria* are open and water is being pumped over the gills, the organism respires aerobically, and the mantle deposits calcium carbonate and organic matrix simultaneously. During periods of shell closure, anaerobic metabolism occurs, and the concentration of succinic acid within the extrapallial fluid (the fluid space between the cells of the outer mantle epithelium and the inner shell surface through which components must pass to be incorporated into the shell)[55] increases. Dissolution of shell calcium carbonate neutralizes the succinic acid produced during anaerobiosis, resulting in an increase in the concentration of calcium ions and succinate in the extrapallial and mantle fluids.[56,57] This carbonate-dissolution process leaves behind a conchiolin residue that forms organic-rich lines in the shell microstructure. With the return to aerobic conditions, shell decalcification ceases and a new phase of shell-building is initiated.[25] From a structural viewpoint, alternating periods of shell secretion and decalcification result in growth increments in the shell of the hard clam.[23,57]

Shell growth proceeds from the beak as an origin, with concentric lines of growth forming on most of the surface of the valves. After the first 24 to 48 hr of life, a shell gland secretes a thin, transparent shell with a smooth surface. This initial shell deposit, prodissoconch I, develops during the earliest part of the straight-hinge veliger stage when the organism possesses a D-shape (Table 1). The prodissoconch I stage persists for approximately 48 hr. During the latter portion of the straight-hinge veliger stage and throughout the duration of the umbone veliger stage, new shell, bearing fine growth striae, is added by the mantle. This secondary larval shell is termed prodissoconch II.[10,58] Shell deposited after the prodissoconch II is called the dissoconch, and it is characterized by well-developed lamellate ridges and concentric lines covered by organic periostracum (a protective organic layer covering the exterior of the shell). Development of the dissoconch shell occurs subsequent to larval settlement and metamorphosis. It marks the initiation of the juvenile stage. The abrupt change in shell morphology from the prodissoconch II to the dissoconch shell appears as a sharp line on the shell surface.

## IV. SHELL MICROSTRUCTURE

The hard-clam shell contains four layers: (1) the inner homogeneous layer; (2) the pallial myostracum (pallial muscle-scar layer); (3) the middle layer; and (4) the outer layer.[13,23] By sectioning valves of specimens perpendicular to the shell surface from the umbo to the ventral margin (Figure 4a) and preparing acetate-peel replicas of the acid-etched, valve cross-sections, shell layers can be studied microscopically. Clark[59] and Kennish et al.[60] have discussed the methods of preparation of the bivalve shell for microscopic examination.

Shell microgrowth patterns are best developed in the outer shell layer. Two categories of shell microgrowth patterns have been identified in valve cross-sections of this species, including cyclical growth patterns and growth breaks (Figure 4b to i).[13,14,16,22] In cross-section, cyclical growth patterns appear as an alternation of calcium carbonate-rich layers and organic-rich lines which constitute growth increments in the microstructure (Figure 4e to h). These patterns originate from alternating periods of shell secretion and dissolution during aerobic and anaerobic respiration, respectively, and their periodicities correlate with various environmental and biological rhythms.[18,23,61] Cyclical growth patterns include sub-daily, daily, bidaily, tidal, lunar-monthly, and annual types (Table 2).

The basic unit of shell growth employed in environmental and physiological studies of the hard clam is the daily growth increment, a calcium carbonate-rich region between successive organic rich lines. The daily growth increment signifies the existence of a circadian rhythm in shell growth of *M. mercenaria* which has been confirmed by mark-and-recapture experiments. For example, Rhoads and Pannella,[14] Cunliffe,[15] and Kennish[20-22] transplanted several thousand hard clams to intertidal and subtidal waters in Connecticut, Massachusetts, and New Jersey. Specimens subsequently recovered and studied contained approximately the same number of growth increments in the outer shell layer as the number of days that had elapsed since the day of transplantation. Pannella and MacClintock[13] kept specimens in tanks for several months without any food, and these clams added a thin growth increment each day. Other specimens placed on a table for 7 days added 7 growth increments.

The character of daily growth increments is strongly influenced by environmental conditions through physiological responses of the bivalve. Water temperature, salinity, tides, water depth, substratum, storms, food supply, spawning events, and the specimen's age affect the thickness of daily growth increments.[12-23] Of these factors, water temperature and age appear to exert the major controlling influence on growth. For instance, the thickness of daily growth increments is greatest in clams between 1 and 4 years of age which are subjected to water temperatures ranging from 20 to 25°C.

Growth breaks result from extremes in environmental conditions, rapid changes in environmental conditions, and physiological stresses which preclude shell secretion for a period of 24 hr or more. At this time, the valves of the organism remain closed, anaerobic respiratory pathways are employed, and shell decalcification takes place.[25,26,56] Some hard clams withdraw their mantle when stressed, producing U- or V-shaped notches in the outer shell layer (Figure 4b, c, i; Figure 5a to e). Occasionally, growth breaks are extensive and can be observed on the external shell surface as linear depressions or concentric rings. Crossed lamellar shell structure and an increased concentration of shell organic material are usually associated with the region of these breaks. Other growth breaks are less extensive because they represent only a few microns of change in the microstructure and are not associated with crossed lamellar shell structure or a high organic content.

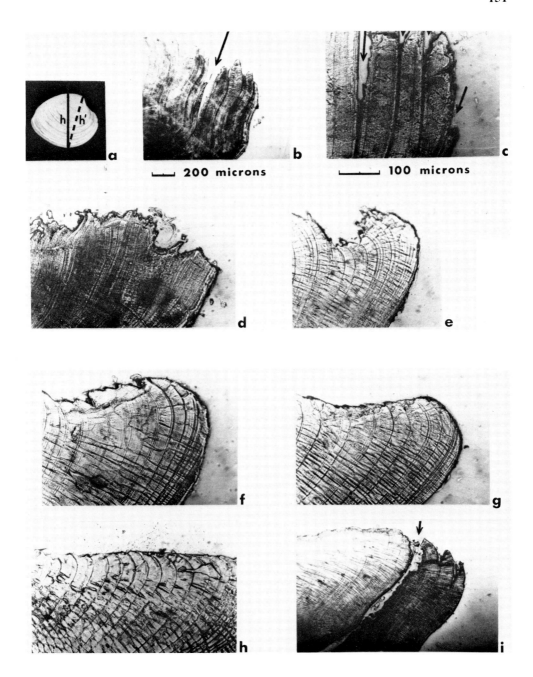

FIGURE 4. (a) Right-hand valve of the hard clam, *M. mercenaria* (Linné). The solid line (H) defines the true height of the organism. The dashed line (h') represents the dimension of height used in this study. (b to d) Optical micrographs of acetate peels showing late winter and early spring microgrowth patterns in the outer shell layer of hard clams from Barnegat Bay, N.J. Shell microgrowth patterns near the edge of the shell include thin daily growth increments, freeze-shock breaks (arrows), and crossed lamellar shell structure (dark areas). (e) Ventral valve margin of a specimen that died in the late spring contains relatively large daily growth increments compared to those of (b to d). Note organic-rich layers (thin, dark bands) bounding calcium-carbonate-rich layers (thick, light layers). (f to h) Summer microgrowth patterns reflect thick daily growth increments characteristic of rapid summer growth in hard clams from temperate waters. Summer microgrowth patterns are distinctive and can be used to determine the absolute age of specimens. (i) Summer storm break (arrow) near the shell margin of *M. mercenaria*. Note the silt grains trapped at the base of the break and the sudden occurrence of the break. The scales for (d to h) equal that of (c). The scale for (i) is equal to that of (b). (From Kennish, M. J., *Environ. Geol.*, 2, 223, 1978. With permission.)

**Table 2**
## CYCLICAL MICROGROWTH PATTERNS IN *MERCENARIA MERCENARIA*

| Pattern | Periodicity (days) | Description |
|---|---|---|
| Subdaily | <1 | This pattern is observed as diffuse organic-rich lines surrounded by calcium carbonate-rich regions. The diffuse organic-rich lines form during intervals of anaerobiosis and shell decalcification within a 24-hr period. The calcium carbonate-rich regions form during episodes of shell secretion. |
| Daily | 1 | A daily growth increment comprises the daily microgrowth pattern. This increment appears as a calcium carbonate-rich region between successive organic-rich lines. In contrast to the diffuse organic-rich lines in the subdaily pattern, diurnal organic-rich lines are well developed. The daily growth increment forms in response to a circadian locomotor rhythm in the species. When the valves of the organism are open and water is freely pumped, aerobic metabolism occurs and calcium carbonate-rich regions of the shell are deposited. When the valves of the clam are closed, anaerobic respiratory pathways are utilized and shell decalcification occurs, resulting in organic-rich growth lines. This alternating sequence of shell deposition and dissolution takes place on a daily basis. |
| Bidaily | 2 | A bidaily growth pattern consists of pairs of growth increments — one thick increment followed by a thin one. The cause of this couplet is unknown. |
| Fortnightly | 14 | Fortnightly patterns appear as a cluster of six to eight thin daily growth increments followed by a cluster of six to eight thick daily growth increments. Alternating neap and spring tides generate this pattern. |
| Lunar-month | 29 | Monthly patterns are clusters of tidal cycles with an overall periodicity of 29 days. This pattern is manifested on the external shell surface as concentric ridges formed by the thickened outer shell layer. The synodic month produces this pattern. |
| Annual | 340—380 | The yearly cycle is composed of a long sequence of thin daily increments followed by a long sequence of thick daily increments. Seasonal temperature changes are mainly responsible for this cyclical pattern, with low winter temperatures yielding thin daily increments and high summer temperatures yielding thick daily increments. |

From Kennish, M. J. and Olsson, R. K., *Environ. Geol.*, 1, 41, 1975. With permission.

Seven types of growth breaks have been identified in the shell of *M. mercenaria:* freeze-shock, heat-shock, thermal-shock, abrasion, spawning, neap-tide, and storm breaks (Table 3). Because each type of growth break is preceded and followed by a unique pattern of daily growth increments and shell structural changes, each one can be identified in the microstructure. Freeze- and heat-shock breaks, caused by temperature stress, generally are the most abundant and conspicuous growth breaks in the shell of this species. Both have similar characteristics even though they develop during different seasons of the year. For example, very thin increments occur prior to and following each break and are associated with a gradual decrease in overall shell convexity. Crossed lamellar shell structure replaces prismatic

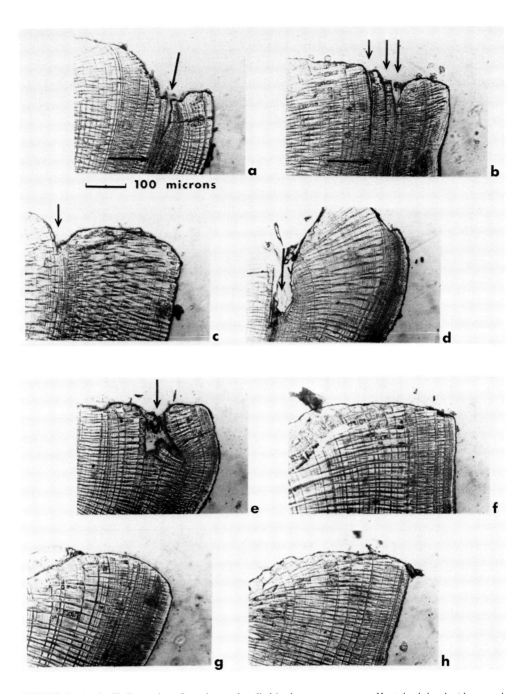

100 microns

FIGURE 5.  (a, b) Shell margins of specimens that died in the summer season. Heat-shock breaks (downward-pointing arrows) and crossed lamellar shell structure border the margins. (c, d) Spawning breaks (arrows) located at the shell edge of hard clams that died in the summer. (e to h) Fall microgrowth patterns that rim the leading edge of the shells of *M. mercenaria*. Note growth increments are of a thickness intermediate between those of the summer (Figure 3f to h) and winter (Figure 3b, c) seasons. Fall microgrowth patterns can be distinguished from spring patterns in the shell microstructure because summer growth patterns exist immediately dorsal to them. A growth break (arrow) occurs in (e). The scales for (b to h) equal that of (a). (From Kennish, M. J., *Environ. Geol.*, 2, 223, 1978. With permission.)

**Table 3**
**GROWTH BREAKS IN *MERCENARIA MERCENARIA***

| Type of break | Description |
|---|---|
| Freeze-shock break | This break is characterized by a deep V-shaped notch in the outer shell layer and an increase in organic material that extends from the outer layer to the inner homogeneous layer. Daily growth increments gradually decrease in thickness as the break is approached and gradually increase in thickness going away from it. Crossed lamellar shell structure replaces prismatic shell structure on one or both sides of the break. The gradual decrease in increment thickness approaching the break reflects the gradual decrease in water temperature associated with the onset of winter. The break itself often marks the first freeze of winter, and the gradual increase of increment thickness following the break indicates the slow recovery from the freeze shock. More than one freeze-shock break may be present per winter season. |
| Heat-shock break | A heat-shock break is structurally similar to a freeze-shock break, but is characteristically developed during the summer months. It apparently occurs when the organism is shocked by excessively high water temperatures. More than one heat-shock break may be present per summer season. |
| Thermal-shock break | A thermal-shock break appears as a sudden break in the normal pattern of growth-increment addition. There is usually no slowdown of growth prior to the break, massive slowdown following the break, and crossed lamellar shell structure throughout the stressed period. This break was first described in specimens from Barnegat Bay, New Jersey, which were affected by rapid fluctuations in water temperatures associated with sharp changes in operations of the Oyster Creek Nuclear Generating Station. |
| Abrasion break | An abrasion break is structurally similar to a thermal-shock break but shows less slowdown of growth following the break. Abrasion breaks are frequently produced by filing or sanding the valve margins during transplantation experiments. |
| Spawning break | In this break, the normal pattern of daily growth-increment addition is abruptly interrupted by a break in growth and immediately followed by a sequence of thin, indistinct increments. The sudden occurrence of the break corresponds to the triggering of the spawning event when the proper temperature is attained. The thin increments associated with the break reflect the period during which the clam stops feeding. Crossed lamellar shell structure is generally absent. |
| Neap-tide break | This break in growth is preceded by a rapid decrease in the thickness of growth increments and is followed by a rapid increase in the thickness of the growth increments. Such a break, when present, coincides with neap tides. Crossed lamellar shell structure is usually absent. |
| Storm break | A storm break appears as a break in the normal depositional pattern of shell growth, which is abruptly followed by recovery. The thickness of growth increments prior to and following the break is not significantly altered. Silt grains are occasionally incorporated into the outer shell layer. Crossed lamellar shell structure is generally lacking. |

From Kennish, M. J. and Olsson, R. K., *Environ. Geol.*, 1, 41, 1975. With permission.

shell structure on one, or each side of the break, and a pronounced U- or V-shaped notch forms in the outer shell layer bordered by a region of high organic content (Figure 4b to d; Figure 5a, b).

A freeze-shock break generally develops during the first freeze of the year although it may also occur later in the winter or early spring. Thin increments preceding a freeze-shock break are indicative of a decrease in shell deposition concomitant with a decline of water temperature in the fall and winter. When water temperature rises, physiological recovery from the period of temperature stress ensues, and growth increments increase in thickness subsequent to the break.

A heat-shock break usually arises during the period of maximum summer temperatures. As water temperature gradually increases in the summer above the optimum for growth in the species, growth increments become progressively thinner until a break in shell deposition is reached. Physiological recovery from the temperature stress results in renewed growth and an increase in the thickness of growth increments after the break.

Because seasonal water temperatures vary from year to year, freeze- and heat-shock breaks rarely develop at precisely the same time each year. However, both types of breaks occur synchronously in a large number of individuals within a given population of hard clams. This synchrony enables investigators to determine the age, growth rate, and mortality patterns of the population.

## V. APPLICATION OF SHELL MICROGROWTH PATTERNS TO AGE DETERMINATIONS OF *MERCENARIA MERCENARIA*

Shell microgrowth patterns can be utilized in age determinations of *M. mercenaria*. For example, counts of daily growth increments from the umbo to the ventral valve margin yield data on absolute age of individual specimens. If these daily increments of growth are measured, daily and yearly growth rates can also be established.

Probably the most accurate method of resolution of absolute age in the hard clam is to combine growth-increment counting with the interpretation of growth breaks in the shell. Periodic annual growth breaks, such as heat- and freeze-shock breaks (Figure 4a, b; Figure 5a, b), are most useful in age determinations. They represent universal, synchronous events in a population;[13,61] thus, by identifying and counting them across the shell, it is possible to obtain the organism's age.

Rhoads and Pannella[14] concluded that the following equation gives an accurate age for some bivalves:

$$Ad = \frac{Ns + Nw}{2} \qquad (2)$$

where Ad is the age of the organism at death, Ns is the number of summer growth breaks, and Nw is the number of winter growth breaks in the shell. This equation may hold for *M. mercenaria* from temperate waters where seasonal changes in water temperature and food supply are substantial, and freeze- and heat-shock breaks tend to be conspicuously developed each year. However, in subtropical waters, the magnitude of seasonal variations in environmental conditions is significantly less than in temperate waters. Consequently, differences between summer and winter microgrowth patterns are smaller and more difficult to detect in subtropical clams. Freeze-shock breaks, in particular, rarely appear in hard clams from subtropical waters. For these bivalves, other periodic annual microgrowth patterns, such as spawning breaks (Figure 5c, d) can be used to determine absolute age.

Because *M. mercenaria* from temperate waters possess well-developed seasonal microgrowth patterns, it is easier to determine their age than specimens from subtropical waters. In temperate hard clams, microgrowth patterns formed during the winter and early spring are characterized by freeze-shock breaks and thin daily increments of growth ($\approx$ 1 to 50 $\mu$m) (Figure 4b to d). In contrast, microgrowth patterns developed in the late spring and summer typically consist of thick daily increments of growth ($\approx$ 50 to 150 $\mu$m) (Figure 4e to h), heat-shock breaks (Figure 5a, b), and spawning breaks (Figure 5c, d). Fall microgrowth patterns are composed of daily growth increments which are intermediate in thickness between those of the winter and summer seasons (Figure 5e to h). In addition, they supercede a long period of summer growth increments, and therefore can be distinguished from other seasonal microgrowth patterns.

Counts of seasonal microgrowth patterns from the umbo to the ventral valve margin will yield an absolute age ( $\pm$ one season) for an individual specimen. Although less accurate than counts of daily growth increments, counts of seasonal microgrowth patterns are especially practical when time constraints place limits on the use of daily growth increments. Counts of daily increments require considerably more analysis time than counts of seasonal microgrowth patterns and may preclude research on large sample sizes.

Approximate ages of specimens can be obtained by the examination of ontogenetic stages recorded in the outer shell layer. Cunliffe[15] discerned three ontogenetic stages in valve cross-sections of this species: (1) youth, (2) maturity, and (3) old age. During the first 2 years of life (youth), growth increments in the outer shell layer intersect the shell surface nearly at right angles, and the outer shell layer terminates in concentric ridges. These concentric ridges disappear and daily growth increments become recurved in shape during the mature stage of life which persists for 3 to 8 years. The final stage of life is old age, and at this time growth increments become thin and perpendicular to the outer shell surface, numerous growth breaks occur, and crossed lamellar shell structure replaces prismatic shell structure in the outer shell layer. Photographic sequences of these ontogenetic stages can be found elsewhere.[15,59,62] Senescence changes in the shell microstructure correlate with gradual diminished metabolism of the hard clam with increasing age[52] and with increased mortality.[22,23]

It is not possible to determine the age of *M. mercenaria* by studying the exterior of the shell. Daily and seasonal growth patterns cannot be resolved accurately on the shell exterior. In addition, many seasonal growth breaks are indistinguishable from disturbance or spawning breaks except when observed in the shell microstructure. Only microscopic investigation of the shell microstructure enables the identification of the diverse origins of shell microgrowth patterns and the establishment of the absolute age of the organism.

## VI. AGE DETERMINATIONS OF *MERCENARIA MERCENARIA* IN BARNEGAT BAY, NEW JERSEY: A TYPE EXAMPLE

Applying the methods of shell microgrowth analysis discussed above, I investigated the age structure of natural populations of *M. mercenaria* in Barnegat Bay, New Jersey, a lagoonal-type estuarine system (Figure 1). During the summer of 1974, death assemblages (assemblages of empty valves)[63] were collected at four sites in the estuary (sites 1 to 4 in Figure 1). Death assemblages give a more realistic picture of the age structure of a population because they contain many generations of individuals.

Specimens were measured and sectioned along an axis from the umbo to the ventral valve margin at a slight angle to the dorsal-ventral plane to investigate size and age relationships (Figure 4a). This axis has been labeled h' to differentiate it from the true shell height (h). A total of 277 valves of hard clams from the four death assemblages was studied microscopically, including 101 from site 1, 74 from site 2, 46 from site 3, and 38 from site 4. Table 4 summarizes the age-h' statistics of these samples. Most of the clams are less than 7 years of age and 70 mm in h' in each assemblage (Table 5). Inspection of the age-h' values for the pooled death assemblages of clams indicates that growth is fastest during the first 4 years of life (Table 6). At this time in ontogeny, specimens add approximately 1 cm of shell each year.

Most *M. mercenaria* in the death assemblages range from 40 to 70 mm in h' corresponding to ages of 4 to 9 years. Thus, mortality is greater in older than younger post-set clams. This mortality pattern is reflected in life tables constructed for each death assemblage based on the absolute age at death of individual specimens (Tables 7 to 10). Since meroplanktonic and pediveliger stages of the hard clam were not examined in this study, the life tables were formulated beginning at age 1 of a hypothetical cohort (radix) of 1000 individuals.

**Table 4**
**AGE-HEIGHT' STATISTICS FOR SAMPLES OF**
**CLAMS FROM SITES 1 TO 4**

| Site | No. of clams | Mean h' (mm)[a] | Standard deviation | Mean age (year) | Standard deviation |
|---|---|---|---|---|---|
| 1 | 101 | 43.12 | 12.81 | 3.92 | 1.70 |
| 2 | 74 | 46.82 | 12.57 | 4.25 | 1.72 |
| 3 | 64 | 49.74 | 9.65 | 5.03 | 1.31 |
| 4 | 38 | 46.82 | 17.05 | 4.03 | 1.79 |

[a] Height' represents a dimension of the shell from the umbo to the ventral margin at a slight angle to the dorsal-ventral plane (see Figure 4a).

From Kennish, M. J., *Environ. Geol.*, 2, 223, 1978. With permission.

**Table 5**
**MEAN AGE-HEIGHT' RELATIONSHIPS OBSERVED IN DEATH**
**ASSEMBLAGES OF CLAMS AT SITES 1 TO 4**

| | Site 1 (N[a] = 101) | | | Site 2 (N[a] = 74) | | | Site 3 (N[a] = 64) | | | Site 4 (N[a] = 38) | | |
|---|---|---|---|---|---|---|---|---|---|---|---|---|
| Age | N[b] | Mean height' (mm)[c] | Std. dev. | N[b] | Mean height' (mm)[c] | Std. dev. | N[b] | Mean height' (mm)[c] | Std. dev. | N[b] | Mean height' (mm)[c] | Std. dev. |
| 1 | 101 | 12.39 | 3.76 | 74 | 11.36 | 4.16 | 64 | 7.04 | 2.71 | 38 | 12.72 | 5.15 |
| 2 | 89 | 23.00 | 4.49 | 68 | 23.51 | 4.96 | 64 | 17.70 | 4.24 | 35 | 25.61 | 6.19 |
| 3 | 65 | 32.91 | 4.49 | 63 | 35.01 | 4.89 | 63 | 29.54 | 5.22 | 25 | 37.30 | 6.42 |
| 4 | 50 | 42.36 | 5.26 | 44 | 43.78 | 4.15 | 54 | 39.55 | 5.23 | 19 | 46.97 | 5.25 |
| 5 | 28 | 49.37 | 6.27 | 19 | 49.89 | 3.01 | 28 | 47.75 | 5.81 | 12 | 55.58 | 7.05 |
| 6 | 16 | 57.61 | 7.76 | 9 | 56.18 | 3.94 | 10 | 55.11 | 4.05 | 6 | 62.13 | 8.88 |
| 7 | 6 | 59.52 | 5.41 | 6 | 63.60 | 1.70 | 7 | 63.30 | 4.56 | 3 | 65.80 | 3.04 |
| 8 | 0 | — | — | 4 | 70.60 | 1.75 | 3 | 69.13 | 5.95 | 0 | — | — |

[a] Number of clams.
[b] Number of size measurements per age.
[c] Height' represents a dimension of the shell from the umbo to the ventral margin at a slight angle to the dorsal-ventral plane (see Figure 4a).

From Kennish, M. J., *Environ. Geol.*, 2, 223, 1978. With permission.

Mortality of *M. mercenaria* is similar at all four locations. Between the ages of 1 and 4 years, hard-clam mortality is low, but it increases significantly through the remaining age intervals. Survivorship curves drawn for populations at each location are strongly convex (Figure 2a to d), indicative of low mortality of individuals until near the end of their lifespan.[64-67] However, survivorship of *M. mercenaria* is very low during the meroplanktonic and prediveliger stages.[8,10,11] Therefore, this species is least capable of coping with physiological and environmental stress during the earliest and latest stages of ontogeny when mortality is greatest. Predation appears to be the factor primarily responsible for low survivorship during the meroplanktonic and pediveliger stages.[10] In the gerontic stage, high mortality seems to be associated with both physiological degeneration[52] and ecological influences.[22]

*M. mercenaria* becomes more susceptible to environmental and physiological perturbations as its metabolism diminishes in the gerontic stage. Heat- and freeze-shocks, major storms, parasitic infestations, and spawning events, for example, can contribute to the increase in

**Table 6**
**MEAN AGE-HEIGHT'**
**RELATIONSHIPS FOR**
**THE POOLED DEATH**
**ASSEMBLAGES OF**
**CLAMS FROM SITES 1**
**TO 4**

| Age | N[a] | Mean height' (mm)[b] | Std. dev. |
|---|---|---|---|
| 1 | 277 | 10.92 | 4.44 |
| 2 | 256 | 22.17 | 5.51 |
| 3 | 216 | 33.05 | 5.67 |
| 4 | 167 | 42.35 | 5.47 |
| 5 | 87 | 49.82 | 6.11 |
| 6 | 41 | 57.35 | 6.65 |
| 7 | 22 | 62.69 | 4.37 |
| 8 | 7 | 69.97 | 3.73 |

*Note:*  The number of clams is 277.

[a]  Number of size measurements per age.
[b]  Height' represents a dimension of the shell from the umbo to the ventral margin at a slight angle to the dorsal-ventral plane (see Figure 4a).

From Kennish, M. J., *Environ. Geol.*, 2, 223, 1978. With permission.

mortality at this time. However, the relative influence of environmental and physiological factors on the overall longevity of this species has not been determined and remains a subject for future research.

Shell microgrowth analysis of natural populations of *M. mercenaria* demonstrates conclusively that this bivalve experiences true senescence. Consequently, it contradicts the view of many investigators that wild populations of molluscs show death well before true aging.[68,69] Studies of the shell microstructure of other bivalves with long lifespans may yield additional examples of wild populations which display true senescence during ontogeny.

## VII. AGE DETERMINATIONS OF OTHER BIVALVES

Shell microgrowth patterns have been used to investigate absolute age in bivalves other than *M. mercenaria.* For example, Rhoads and Pannella[14] determined the absolute age of *Gemma gemma* in Long Island Sound by counting daily growth increments in dead specimens. Tevesz[70] also examined the absolute age of *G. gemma;* however, he used counts of tidal microgrowth patterns rather than daily growth increments. Jones et al.,[71] Thompson,[72] and Ropes and O'Brien[73] studied the age structures of commercially important, shelf-clam species, the ocean quahog (*Arctica islandica*) and the surf clam (*Spisula solidissima.*) Counts of annual growth bands in the shell microstructure provided absolute age measurements of individual specimens. Results of Jones et al.[71] and Thompson[72] indicate that most surf clams

## Table 7
### LIFE TABLE FOR *MERCENARIA MERCENARIA* AT SITE 1

| Age interval x to x + 1 years | Proportion dying in interval (x, x + 1) 1000 $q_x$ | Number living at age x $l_x$ | Number dying in interval (x, x + 1) $d_x$ | Number of time-spans lived in interval (x, x + 1) $L_x$ | Total number of time-spans lived past age x $T_x$ | Average life expectancy (in years) at age x $e_x$ | Proportion surviving in interval (x, x + 1) $P_x$ |
|---|---|---|---|---|---|---|---|
| 1—2 | 22.00 | 1000 | 22 | 989.0 | 4508.0 | 4.5080 | 0.9780 |
| 2—3 | 68.51 | 978 | 67 | 944.5 | 3519.0 | 3.5982 | 0.9315 |
| 3—4 | 83.42 | 911 | 76 | 873.0 | 2574.5 | 2.8260 | 0.9166 |
| 4—5 | 130.54 | 835 | 109 | 780.5 | 1701.5 | 2.0377 | 0.8695 |
| 5—6 | 544.08 | 726 | 395 | 528.5 | 921.0 | 1.2686 | 0.4559 |
| 6—7 | 314.20 | 331 | 104 | 279.0 | 392.5 | 1.1858 | 0.6858 |
| 7—8 | 1000.00 | 227 | 227 | 113.5 | 113.5 | 0.5000 | 0.0000 |

From Kennish, M. J., *Environ. Geol.*, 2, 223, 1978. With permission.

## Table 8
### LIFE TABLE FOR *MERCENARIA MERCENARIA* AT SITE 2

| Age interval x to x + 1 years | Proportion dying in interval (x, x + 1) 1000 $q_x$ | Number living at age x $l_x$ | Number dying in interval (x, x + 1) $d_x$ | Number of time-spans lived in interval (x, x + 1) $L_x$ | Total number of time-spans lived past age x $T_x$ | Average life expectancy (in years) at age x $e_x$ | Proportion surviving in interval (x, x + 1) $P_x$ |
|---|---|---|---|---|---|---|---|
| 1—2 | 7.00 | 1000 | 7 | 996.5 | 4844.0 | 4.8440 | 0.9930 |
| 2—3 | 27.19 | 993 | 27 | 979.5 | 3847.5 | 3.8746 | 0.9728 |
| 3—4 | 39.34 | 966 | 38 | 947.0 | 2868.0 | 2.9689 | 0.9607 |
| 4—5 | 113.15 | 928 | 105 | 875.5 | 1921.0 | 2.0700 | 0.8869 |
| 5—6 | 343.86 | 823 | 283 | 681.5 | 1045.5 | 1.2704 | 0.6561 |
| 6—7 | 829.63 | 540 | 448 | 316.0 | 364.0 | 0.6741 | 0.1704 |
| 7—8 | 978.26 | 92 | 90 | 47.0 | 48.0 | 0.5217 | 0.0217 |
| 8—9 | 1000.00 | 2 | 2 | 1.0 | 1.0 | 0.5000 | 0.0000 |

From Kennish, M. J., *Environ. Geol.*, 2, 223, 1978. With permission.

**Table 9**
**LIFE TABLE FOR *MERCENARIA MERCENARIA* AT SITE 3**

| Age interval x to x + 1 years | Proportion dying in interval (x, x + 1) 1000 $q_x$ | Number living at age x $l_x$ | Number dying in interval (x, x + 1) $d_x$ | Number of time-spans lived in interval (x, x + 1) $L_x$ | Total number of time-spans lived past age x $T_x$ | Average life expectancy (in years) at age x $e_x$ | Proportion surviving in interval (x, x + 1) $P_x$ |
|---|---|---|---|---|---|---|---|
| 1—2 | 0.00 | 1000 | 0 | 1000.0 | 5074.0 | 5.0740 | 1.0000 |
| 2—3 | 0.00 | 1000 | 0 | 1000.0 | 4074.0 | 4.0740 | 1.0000 |
| 3—4 | 30.00 | 1000 | 30 | 985.0 | 3074.0 | 3.0740 | 0.9700 |
| 4—5 | 122.68 | 970 | 119 | 910.5 | 2089.0 | 2.1536 | 0.8773 |
| 5—6 | 324.32 | 851 | 276 | 713.0 | 1178.5 | 1.3848 | 0.6757 |
| 6—7 | 775.65 | 575 | 446 | 352.0 | 465.5 | 0.8096 | 0.2244 |
| 7—8 | 620.16 | 129 | 80 | 89.0 | 113.5 | 0.8798 | 0.3798 |
| 8—9 | 1000.00 | 49 | 49 | 24.5 | 24.5 | 0.5000 | 0.0000 |

From Kennish, M. J., *Environ. Geol.*, 2, 223, 1978. With permission.

**Table 10**
**LIFE TABLE FOR *MERCENARIA MERCENARIA* AT SITE 4**

| Age interval x to x + 1 years | Proportion dying in interval (x, x + 1) 1000 $q_x$ | Number living at age x $l_x$ | Number dying in interval (x, x + 1) $d_x$ | Number of time-spans lived in interval (x, x + 1) $L_x$ | Total number of time-spans lived past age x $T_x$ | Average life expectancy (in years) at age x $e_x$ | Proportion surviving in interval (x, x + 1) $P_x$ |
|---|---|---|---|---|---|---|---|
| 1—2 | 19.00 | 1000 | 19 | 990.5 | 4951.0 | 4.9510 | 0.9810 |
| 2—3 | 34.66 | 981 | 34 | 964.0 | 3960.5 | 4.0372 | 0.9653 |
| 3—4 | 24.29 | 947 | 23 | 935.5 | 2996.5 | 3.1642 | 0.9757 |
| 4—5 | 124.46 | 924 | 115 | 866.5 | 2061.0 | 2.2305 | 0.8755 |
| 5—6 | 398.02 | 809 | 322 | 648.0 | 1194.5 | 1.4765 | 0.6020 |
| 6—7 | 377.82 | 487 | 184 | 395.0 | 546.5 | 1.1222 | 0.6222 |
| 7—8 | 1000.00 | 303 | 303 | 151.5 | 151.5 | 0.5000 | 0.0000 |

From Kennish. M. J., *Environ. Geol.*, 2, 223, 1978. With permission.

live less than 20 years, whereas most ocean quahogs live an average of approximately 50 years. These data are critical to resource management programs assessing these commercially important molluscs.

Other bivalves possess shell microgrowth patterns which have great potential application to age determination studies. For instance, some bivalves have a daily periodicity of shell growth; these include *Argopecten irradians*,[74] *Cardium* spp.,[75] *Mactra* and *Chione* spp.,[12] *Cardium edula*,[76] *Pecten diegensis*,[77] *Tridacna squamosa*,[13] and *Clinocardium nuttalli*.[78] Analysis of daily growth increments in these bivalves may prove valuable to research on longevity and other aspects of life history.

The documentation of the periodicities of microgrowth patterns in bivalve shells is essential to research on age and senescence. Probably the most effective method of establishing such periodicities is to undertake mark-and-recapture experiments with individuals of many age groups over sufficiently long periods time.[22,61] In addition, shell microgrowth studies conducted on experimental animals raised under controlled laboratory conditions, utilizing recently developed culture techniques, could be most informative. Once periodicities of shell microgrowth patterns are established, applications to practical problems such as research in population dynamics and gerontology,[23] can proceed more rapidly.

The bivalve shell is considered to be a continuous physiological and environmental recorder.[14,22] At present, the application of shell microgrowth patterns to age determinations of bivalves has been limited to only a few species. However, similar applications to other molluscan taxa should be forthcoming as the relationship of shell microgrowth patterns to specific environmental parameters and physiological conditions become established for these taxa by future research.

## ACKNOWLEDGMENTS

The research reported in this chapter was funded by the Geological Society of America, the Society of Sigma Xi, the Marine Sciences Center of Rutgers University, and the Jersey Central Power & Light Company.

## REFERENCES

1. **Marteil, L.,** Acclimation du clam (*Venus mercenaria L.*) en Bretagne, *Revue Trav. Inst. Peches. Marit.,* 20, 157, 1956.
2. **Heppell, D.,** The naturalization in Europe of the quahog *Mercenaria mercenaria* (L.), *J. Conchol.,* 25, 21, 1961.
3. **Ansell, A. D.,** *Venus mercenaria* in Southampton water, *Ecology,* 44, 396, 1964.
4. **Ansell, A. D.,** The rate of growth of the hard clam *Mercenaria mercenaria* (L) throughout the geographical range, *J. Cons., Cons. Int. Explor. Mer,* 31, 364, 1968.
5. **Castagna, M. and Chanley P.,** Salinity tolerance of some marine bivalves from estuarine environments in Virginia waters on the western mid-Atlantic coast, *Malacologia,* 12, 47, 1973.
6. **Maurer, D., Watling, L., and Aprill, G.,** The distribution and ecology of common marine and estuarine pelecypods in the Delaware Bay area, *Nautilus,* 88, 38, 1974.
7. **McHugh, J. L.,** Estuarine fisheries: are they doomed?, in *Estuarine Processes,* Vol. 1, Wiley, M. L., Ed., Academic Press, N.Y., 1976, 15.

8. **MacKenzie, C. L.,** Management for increasing clam abundance, *Mar. Fish. Rev.,* 41, 10, 1979.

9. **Loosanoff, V. L. and Davis, H. C.,** Rearing of bivalve mollusks, in *Advances in Marine Biology,* Vol. 1, Russell, F. S., Ed., Academic Press, N.Y., 1963, 1.

10. **Carriker, M. R.,** Interrelation of functional morphology, behavior and autecology in early stages of the bivalve *Mercenaria mercenaria, J. Elisha Mitchell Sci. Soc.,* 77, 168, 1961.

11. **MacKenzie, C. L.,** Predation on hard clam (*Mercenaria mercenaria*) populations, *Trans. Am. Fish. Soc.,* 106, 530, 1977.

12. **Barker, R. M.,** Microtextural variations in pelecypod shells, *Malacologia,* 2, 69, 1964.

13. **Pannella, G. and MacClintock, C.,** Biological and environmental rhythms reflected in molluscan shell growth, In Paleobiological Aspects of Growth and Development: a Symposium, Macurda, D. B., Ed., Paleontol. Soc. Mem. 2 *(J. Paleontol.),* Suppl. 42, 64, 1968.

14. **Rhoads, D. C. and Pannella, G.,** The use of molluscan shell growth patterns in ecology and paleocology, *Lethaia,* 3, 143, 1970.

15. **Cunliffe, J. E.,** Description, Interpretation, and Preservation of Growth Increment Patterns in Shells of Cenozoic Bivalves, Ph.D. thesis, Rutgers University, New Brunswick, N.J., 1974.

16. **Cunliffe, J. E. and Kennish, M. J.,** Shell growth patterns in the hard-shelled clam, *Underwater Nat.,* 8, 20, 1974.

17. **Kennish, M. J. and Olsson, R. K.,** Effects of thermal discharges on the microstructural growth of *Mercenaria mercenaria, Environ. Geol.,* 1, 41, 1975.

18. **Thompson, I.,** Biological clocks and shell growth in bivalves, in *Growth Rhythms and the History of the Earth's Rotation,* Rosenberg, G. D. and Runcorn, S. K., Eds., John Wiley & Sons, London, 1975, 149.

19. **Kennish, M. J.,** Monitoring thermal discharges: a natural method, *Underwater Nat.,* 9, 8, 1976.

20. **Kennish, M. J.,** Growth increment analysis of *Mercenaria mercenaria* from artificially heated coastal marine waters: a practical monitoring method, *Proc. XII Int. Soc. Chronobiol. Conf.,* 663, 1977.

21. **Kennish, M. J.,** Effects of Thermal Discharges on Mortality of *Mercenaria mercenaria* in Barnegat Bay, New Jersey, Ph.D. thesis, Rutgers University, New Brunswick, N.J., 1977.

22. **Kennish, M. J.,** Effects of thermal discharges on mortality of *Mercenaria mercenaria* in Barnegat Bay, New Jersey, *Environ. Geol.,* 2, 223, 1978.

23. **Kennish, M. J.,** Shell microgrowth analysis: *Mercenaria mercenaria* as a type example for research in population dynamics, in *Skeletal Growth of Aquatic Organisms,* Rhoads, D. C. and Lutz, R. A., Eds., Plenum Press, N.Y., 1980, 255.

24. **Kennish, M. J. and Loveland, R. E.,** Growth models of the northern quahog, *Mercenaria mercenaria* (Linné), *Proc. Natl. Shellfish. Assoc.,* 70, 230, 1980.

25. **Lutz, R. A. and Rhoads, D. C.,** Anaerobiosis and a theory of growth line formation, *Science,* 198, 1222, 1977.

26. **Gordon, J. and Carriker, M. R.,** Growth lines in a bivalve mollusk: subdaily patterns and dissolution of the shell, *Science,* 202, 519, 1978.

27. **Gordon, J.,** Evidence for sclerotization of the shell matrix of a marine bivalve, *Proc. Natl. Shellfish. Assoc.,* 70, 125, 1980.

28. **Dalton, R. C.,** The Reproductive Cycles of the Northern and Southern Quahogs *Mercenaria mercenaria* (Linné) and *Mercenaria campechiensis* (Gmelin) and their Hybrids, with a Note on their Growth, Ph.D. thesis, Florida State University, Tallahassee, 1977.

29. **Loosanoff, V. L.,** Sexual phases in the quahog, *Science,* 83, 287, 1936.

30. **Loosanoff, V. L.,** Development of the primary gonad and sexual phases in *Venus mercenaria* Linnaeus, *Biol. Bull.,* 72, 389, 1937.

31. **Saila, S. and Pratt, S. D.,** Mid-Atlantic Bight fisheries, in *Coastal and Offshore Environmental Inventory, Cape Hatteras to Nantucket Shoals,* Mar. Publ. Ser. No. 2, Universiy of Rhode Island, Kingston, 6-1, 1973.

32. **Eversole, A. G., Michener, W. K., and Eldridge, P. J.,** Reproductive cycle of *Mercenaria mercenaria* in a South Carolina estuary, *Proc. Natl. Shellfish. Assoc.,* 70, 22, 1980.

33. **Nelson, T. C. and Haskin, H. H.,** On the spawning behavior of the oyster and of *Venus mercenaria* with especial reference to the effects of spermatic hormones, *Anat. Rec.,* 105, 484, 1949.

34. **Davis, H. C. and Chanley, P. E.,** Spawning and egg production of oysters and clams, *Biol. Bull.,* 110, 117, 1956.

35. **Bricelj, V. M. and Malouf, R. E.,** Aspects of reproduction of hard clams (*Mercenaria mercenaria*) in Great South Bay, New York, *Proc. Natl. Shellfish. Assoc.,* 70, 216, 1980.

36. **Chanley, P. E. and Andrews, J. D.,** Aids for identification of bivalve larvae of Virginia, *Malacologia,* 11, 45, 1971.

37. **Turner, H. J. and George, C. J.,** Some aspects of the behavior of the quahaug, *Venus mercenaria,* during the early stages, *8th Rep. Invest. Shellfish. Mass.,* Dept. Natl. Res., Div. Mar. Fish., Common. Mass., 1955, 5.

38. **Davis, H. C. and Guillard, R. R.,** Relative value of ten genera of microorganisms as foods for oyster and clam larvae, *Fish. Bull., U.S.,* 58, 293, 1958.
39. **Thorson, G.,** Reproductive and larval ecology of marine bottom invertebrates, *Biol. Rev.,* 25, 1, 1950.
40. **Kennedy, V. S., Roosenburg, W. H., Castagna, M., and Mihursky, J. A.,** *Mercenaria mercenaria* (Mollusca: Bivalvia): temperature-time relationships for survival of embryos and larvae, *Fish. Bull., U.S.,* 72, 1160, 1974.
41. **Chanley, P. E.,** Possible causes of growth variations in clam larvae, *Proc. Natl. Shellfish. Assoc.,* 45, 84, 1955.
42. **Loosanoff, V. L., Miller, W. S., and Smith, P. B.,** Growth and setting of larvae of *Venus mercenaria* in relation to temperature, *J. Mar. Res.,* 10, 59, 1951.
43. **Thorson, G.,** Some factors influencing the recruitment and establishment of marine benthic communities, *Neth. J. Sea Res.,* 3, 267, 1966.
44. **Keck, R., Maurer, D., and Malouf, R.,** Factors influencing the setting behavior of larval hard clams, *Mercenaria mercenaria, Proc. Natl. Shellfish. Assoc.,* 64, 59, 1974.
45. **Pratt, D. M.,** Abundance and growth of *Venus mercenaria* and *Callocardia morrhua* in relation to the character of the bottom sediments, *J. Mar. Res.,* 12, 60, 1953.
46. **Pratt, D. M. and Campbell, D. A.,** Environmental factors affecting growth in *Venus mercenaria, Limnol. Oceanogr.,* 1, 2, 1956.
47. **Carriker, M. R.,** The role of physical and biological factors in the culture of *Crassostrea* and *Mercenaria* in a salt pond, *Ecol. Monogr.,* 29, 221, 1959.
48. **Davis, H. C. and Calabrese, A.,** Combined effects of temperature and salinity on development of eggs and growth of larvae of *M. mercenaria* and *C. virginica, Fish. Bull., U.S.,* 63, 643, 1964.
49. **Turner, H. J.,** A review of the biology of some commercial molluscs of the east coast of North America, in *Sixth Report on Investigations of the Shellfisheries of Massachusetts,* Dept. Natl. Res., Div. Mar. Fish., Common. Mass., 1953, 39.
50. **Henderson, J. T.,** Lethal temperatures of lamellibranchiata, *Contrib. Can. Biol. Fish.,* 4, 399, 1929.
51. **Castagna, M. and Kraeuter, J. N.,** Manual for growing the hard clam *Mercenaria,* Spec. Rep. Appl. Mar. Sci. Ocean Eng. No. 249, Sea Grant Program, Virginia Institute of Marine Science, Wachapreague, 1981.
52. **Hopkins, H. S.,** Age difference and the respiration in muscle tissues of molluscs, *J. Exp. Zool.,* 56, 209, 1930.
53. **Belding, D. L.,** The quahaug fishery of Massachusetts, Dept. Conserv., Div. Fish. Game, Common. Mass., Mar. Fish. Ser. 2, 1931.
54. **Seed, R.,** Shell growth and form in the Bivalvia, in *Skeletal Growth of Aquatic Organisms,* Rhoads, D. C. and Lutz, R. A., Eds., Plenum Press, N.Y., 1980, 23.
55. **Crenshaw, M. A.,** Mechanisms of shell formation and dissolution, in *Skeletal Growth of Aquatic Organisms,* Rhoads, D. C. and Lutz, R. A., Eds., Plenum Press, N.Y., 1980, 115.
56. **Crenshaw, M. A. and Neff, J. M.,** Decalcification at the mantle-shell interface in molluscs, *Am. Zool.,* 9, 881, 1969.
57. **Lutz, R. A. and Rhoads, D. C., Eds.,** Growth patterns within the molluscan shell: an overview, in *Skeletal Growth of Aquatic Organisms,* Plenum Press, N.Y., 1980, 203.
58. **Carriker, M. R. and Palmer, R. E.,** Ultrastructural morphogenesis of prodissoconch and early dissoconch valves of the oyster *Crassostrea virginica, Proc. Natl. Shellfish. Assoc.,* 69, 103, 1979.
59. **Clark, G. R., II,** Study of molluscan shell structures and growth lines using thin sections, in *Skeletal Growth of Aquatic Organisms,* Rhoads, D. C. and Lutz, R. A., Eds., Plenum Press, N.Y., 1980, 603.
60. **Kennish, M. J., Lutz, R. A., and Rhoads, D. C.,** Preparation of acetate peels and fractured sections for observation of growth patterns within the bivalve shell, in *Skeletal Growth of Aquatic Organisms,* Rhoads, D. C. and Lutz, R. A., Eds., Plenum Press, N.Y., 1980, 597.
61. **Clark, G. R., II,** Growth lines in invertebrate skeletons, *Annu. Rev. Earth Planet. Sci.,* 2, 77, 1974.
62. **Clark, G. R., II,** Seasonal growth variations in the shells of recent and prehistoric specimens of *Mercenaria mercenaria* from St. Catherines Island, Georgia, *Am. Mus. Nat. Hist. Anthropol. Pap.,* 56, 161, 1979.
63. **Johnson, R. G.,** Models and methods for analysis of the mode of formation of fossil assemblages, *Geol. Soc. Am. Bull.,* 71, 1075, 1960.
64. **Deevey, E. S.,** Life tables for natural populations of animals, *Q. Rev. Biol.,* 22, 283, 1947.
65. **Odum, E. P.,** *Fundamentals of Ecology,* 3rd ed., W. B. Saunders, Philadelphia, 1971.
66. **Cerrato, R. M.,** Demographic analysis of bivalve populations, in *Skeletal Growth of Aquatic Organisms,* Rhoads, D. C. and Lutz, R. A., Eds., Plenum Press, N.Y., 1980, 417.
67. **Levinton, J. S.,** *Marine Ecology,* Prentice-Hall, Englewood Cliffs, N.J., 1982.
68. **Comfort, A.,** The duration of life in molluscs, *Proc. Malacol. Soc. London,* 52, 219, 1957.
69. **Wilbur, K. M. and Owen, G.,** Growth, in *Physiology of Mollusca,* Vol. 1, Wilbur, K. M. and Yonge, C. M., Eds., Academic Press, London, 1964, 211.

70. **Tevesz, M. J. S.,** Implications of absolute age and season of death data compiled for Recent *Gemma gemma, Lethaia,* 5, 41, 1972.

71. **Jones, D. S., Thompson, I., and Ambrose, W.,** Age and growth rate determinations for the Atlantic surf clam *Spisula solidissima,* (Bivalvia: Mactracea), based on internal growth lines in shell cross-sections, *Mar. Biol.,* 47, 63, 1978.

72. **Thompson, I.,** Growth rates, age composition, and natural mortality of the ocean quahog and surf clam: commercially important New Jersey shelf bivalves, N.J. Mar. Sci. Consort., Sea Grant Program, Sandy Hook, N.J., 1979, 3.

73. **Ropes, J. W. and O'Brien, L.,** A unique method of aging surf clams, *Bull. Am. Malacol. Union,* p. 58, 1979.

74. **Davenport, C. B.,** Growth lines in fossil pectens as indicators of past climates, *J. Paleontol.,* 12, 514, 1938.

75. **Peterson, G. H.,** Notes on the growth and biology of the different *Cardium* species in Danish brackish water areas, *Medd. Dan. Fisk. -og Havunders.,* 2, 1, 1958.

76. **House, M. R. and Farrow, G. E.,** Daily growth banding in the shell of the cockle, *Cardium edule, Nature (London),* 219, 1384, 1968.

77. **Clark, G. R., II,** Mollusk shell: daily growth lines, *Science,* 161, 800, 1968.

78. **Evans, J. W.,** Tidal growth increments in the cockle *Clinocardium nuttalli, Science,* 176, 416, 1972.

Chapter 8

# MARINE MUSSELS AND CEPHALOPODS AS MODELS FOR STUDY OF NEURONAL AGING

**Maynard H. Makman and George B. Stefano**

## TABLE OF CONTENTS

# I. INTRODUCTION

The phylum *Mollusca* includes a diverse group of animals, varying greatly in size, complexity, habitat, life span, and in other respects as well. In this chapter we will consider two marine molluscs, *Mytilus edulis* and *Octopus bimaculatus* as providing, respectively, a relatively simple and a rather complex nervous system for examination in studies of neuronal aging. These species are described in some detail here largely because of our own familiarity with them. Other similar species (as well as other types) of molluscs may of course provide equally good and interesting models for studies of aging and also of development and maturation. Thus, it is hoped that the information concerning *Octopus* neurobiology presented in this chapter, derived primarily from studies of *O. vulgaris* or our own work with *O. bimaculatus*, will be applicable for the most part to investigations of other species of *Octopus*.

# II. GENERAL CONSIDERATIONS FOR STUDY OF NEURONAL AGING IN *OCTOPUS* AND *MYTILUS EDULIS*

## A. *Octopus*

Most notably due to the work of Young[1-2] and Wells[3] considerable information is available concerning the nervous system and the behavior, as well as many other aspects of the biology of *Octopus*. The studies of these and at least until recently, of other investigators have concerned primarily *O. vulgaris*. There are over 100 known species in the genus *Octopus*.[4] We have worked primarily with *O. bimaculatus*, a small, relatively common two-spotted octopus found along the rocky shores of southern California. Another species, *O. bimaculoides* may be found in the same area and is very similar to *O. bimaculatus* in appearance. *Octopuses* may be classified according to the size of their eggs and the nature of the hatchlings. The animals may be divided into two types of species: (1) those including *O. bimaculoides*, with large eggs and relatively adult-like benthonic hatchlings, and (2) those, including *O. bimaculatus*, with a considerably greater number of small, less yolky eggs and small planktonic hatchlings. It has been possible to rear several of the large egg species, while none of the small egg species have been reared as yet.[5-8]

*O. bimaculatus* ranging from about 50 to 500 g in weight have been obtained from Pacific Bio-Marine Laboratories and kept for study in artificial sea water (Instant Ocean) at 13 to 15°C, pH 7.6 to 7.8 specific gravity 1.026, in a closed system. The animals in the wild are solitary except for a brief time during the mating season. Also, they are nocturnal animals. In the laboratory, in order to avoid fighting among the animals, they are housed individually, either in separate tanks or in a common tank with partitions separating the animals from one another. A dark glass jar or clay pot affords a suitable hiding or resting place in the aquarium. The animals are generally fed mussels or crabs. They may be given food that has been stored frozen but prefer live prey.

The lifespan of octopuses in general appears to be about 1.5 to 2 years[3] (probably less than 2 years) and this also appears to be the case for *O. bimaculatus*. Female octopuses generally die within a few weeks after spawning. This may be considered to be a "programmed death," controlled by secretions of the optic gland.[9] While it is not clear whether or not a comparable phenomenon occurs for males, male and female animals appear to have similar lifespans. (See Section II.C. below for further discussion.) Growth and lifespan of several species of *Octopus* have been studied,[3] but detailed information is not available for *O. bimaculatus*. Both young and mature adult animals, including females about to or that have spawned, are available commercially.

## B. *Mytilus edulis*

Bivalve growth has been shown to vary with regard to age and environmental conditions (for review, see Seed[10]). *M. edulis* previously employed in our studies were subtidal and harvested from Long Island Sound at Northport, N.Y. They can be found all along the northeast coast. In using this particular area for the past 7 years, representative populations of *M. edulis* have been monitored for aging studies. Ultimately, the most valid study of an animal is that involving examination of the animal in its own environment. However, maintaining an animal in the laboratory is not only important, but also may be necessary for controlled study and evaluation of various parameters. It should be kept in mind that in the laboratory setting, the degree of interaction with a challenging environment is greatly reduced and this may ultimately be reflected in the animal's reactive capacity both physiologically and biochemically.

Maintaining colonies of *M. edulis* in an open environment can be relatively easy if certain precautions are taken from the start. Animals which are 0.5 cm long can be gathered subtidally or from intertidal regions where they can be found on rocks or dock pilings. They can then be placed on subtidal rocks previously cleaned by scraping. Large numbers of these animals can be transplanted successfully by pinning them down to the rocks by placing a nylon mesh net (0.25 cm openings) over them. The net is weighed on the perimeter to prevent it from moving. After about 3 days, animals which attached themselves to the net, not the rock, are removed from the net and the process is repeated until nearly all the animals attached themselves to the rock. At this time the net fitting is loosened and just serves as a means of minimizing predation. In this way many colonies can be started in the same immediate area. In subsequent years younger animals (smaller) can be easily removed. The growth rates obtained over a 4-year period appear to be quite uniform.[11-13] In this subtidal locale the mussels grow approximately 5.0 cm in the first 2 years (starting from 0.5 cm), then 1.0 for the third and about 0.8 for the fourth (Figure 1). By the fifth year, in this area, the animals only grow another 0.5 cm, which is really negligible considering individual variations.[12] In some areas *M. edulis* has been reported to live to 16 years of age.[10] Intertidal *M. edulis* tend not to grow as well as their subtidal counterparts,[10,12] again indicating the importance of the immediate environment of the animals (Figure 1).

Maintaining *M. edulis* in the laboratory is relatively easy. Whether obtained commercially (scientific supplier) or by the investigator directly from the marine environment, the animals do well in the laboratory setting. However, animals harvested in the winter do not adapt well to the laboratory. This is probably related to thermal shock and even trying to acclimate the animals for short periods of time does not work well. If one obtains animals from a fish market any time of the year, there will also be a high mortality rate. In the laboratory animals can be kept in aerated artificial seawater (Instant Ocean) at 15 to 19°C, pH 7.5 to 7.8, and specific gravity 1.025.

In selecting healthy animals for experimentation three physiological criteria are used:

1.  Shell closes quickly when the siphon is touched.
2.  Touching the foot of a gaping animal causes its withdrawal.
3.  The valves remain closed against a reasonable amount of force applied to open them.

In addition, upon opening the animal, the visible tissues are assessed for the presence of growths, tissue damage, or abnormal discolorations.

Animals in the laboratory setting do not grow at the rate of those found in their natural environment. In the first 2 years animals grew from 0.5 cm to 2.1 cm (unpublished data). These laboratory animals were maintained on filtered seawater which was changed every other day.

## C. Hormonal Inhibition of Feeding and Programmed Death in *Octopus:* Role of the Optic Gland

The optic glands of *Octopus* are known to be involved in regulation of gonadal maturation and reproduction.[3] In addition, these glands appear to function to inhibit feeding and thereby

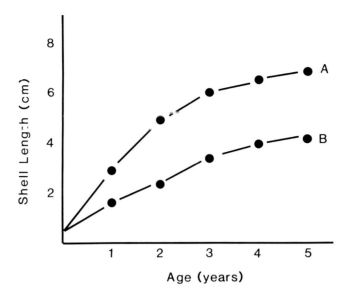

FIGURE 1.    Growth curves of subtidal (A) and intertidal (B) *M. edulis*
maintained *in situ* at Northport, Long Island.[12]

to cause the death of the animal. The relationship of gonadal function and feeding has been
examined in some detail in *O. hummelincki*. Female *O. hummelincki* spawn only once in
their lives, as in the case of *Octopus* in general. After spawning the animals eat less while
caring for the eggs and they die shortly after the eggs hatch (about 30 to 50 days after
egglaying). Females with optic glands removed bilaterally cease to incubate the eggs and
resume a feeding pattern similar to that present before egglaying. After a transient period
of weight loss (probably related to the egglaying itself) the females gradually increase their
body weight, attaining weights above the maximum typical of the species, and live much
longer (time from egglaying to death 75 to 277 days).[9] Males without optic glands also
surpass the expected maximum body weight and appear to live longer than normal males
(although there is no fixed reference point comparable to egglaying in the female to accurately
evaluate change in lifespan.) In the male, the linkage of inhibition-of-feeding and death to
reproductive activities seemed to be looser than in the female, but still present. It has been
postulated that in the female this mechanism may help to ensure survival of the eggs by
preventing the female from eating them.[9] Whether or not regulatory mechanisms involving
the optic gland other than or in addition to cessation of feeding are responsible for reduced
longevity has not yet been established. Also, the humoral factors in the optic gland mediating
these effects remain to be isolated and characterized.

## III. NERVOUS SYSTEM: ANATOMY AND GENERAL FEATURES

### A. *Octopus*

The octopods have a highly developed and complex nervous system and are capable of
complicated patterns of behavior.[1,3,15] For an outline of the general anatomy of *Octopus,*
including external features, layout of internal organs, and descriptions of digestive, respiratory, circulatory, excretory, reproductive, and endocrine systems as well as of sensory
organs, the reader is referred to the review of Wells.[3] On the basis of total weight and the
number of neurons, the *Octopus* nervous system indeed equals or exceeds that of many birds
and mammals. The brain-to-body weight ratio alone exceeds that of most fish and reptiles.[3]
An *O. bimaculatus* with body weight 300 to 500 g will have a main central mass of brain

tissue (supra- and subesophageal lobes) weighing over 100 mg and optic lobes weighing perhaps over 200 mg. The brain weight of a 1-kg *Octopus* is about 1 g.[3] The CNS of *Octopus* and other cephalopods is formed from a system of cords rather than concentrations of distinct ganglia.[1] Compared with the vertebrate nervous system, that of *Octopus* is very diffuse with most of the neurons outside the brain. The nerve cords of the arms alone contain $3.5 \times 10^8$ cells, more than twice the number in the main central mass of brain (about $0.5 \times 10^8$ cells) plus the optic lobes (about $1 \times 10^8$ cells).[2,16] The rest of the nervous system contains about $1.5 \times 10^8$ cells, although this is only a very rough estimate. The CNS of *Octopus* is made up of a large number of small neurons with many interconnecting pathways. The optic lobe and the higher centers of brain such as the vertical lobe, in particular, contain a concentration of small neurons. The optic lobe itself has a highly organized structure consisting of many layers and many different neuronal cell types. In this chapter we will refer to the supra- and subesophageal lobes together as the "brain." It should be noted however, that the optic lobes are really an integral part of that same central structure and thus could also be considered as part of the "brain". Our discussion in this Chapter will be concerned primarily with the brain and optic lobes, the neural structures most responsible for higher integrative functions in *Octopus*.

The general anatomical relationships of the major lobes of brain, optic lobes, and eyes are shown in Figures 2 and 3. Many of the complex interconnections of these various structures in *Octopus* have been summarized by Young,[1] and this investigator has also more recently carried out extensive studies of the organization of the squid nervous system.[17-20] The horseradish peroxidase and cobalt chloride injection methods have been utilized in a number of recent studies of *Octopus* peripheral and central nervous system, including examination of the connections of the optic lobe.[21-26] Since the brain of octopus serves to integrate a variety of functions (sensory, motor, cardiovascular, respiratory, etc.), it is difficult to ascribe specific functions or processes to a single brain region. Motor control may be considered as involving the following systems:[1] (1) lower motor centers, e.g., ganglia in the arms (influenced by central regulatory processes and also capable of direct reflex action due to impulses from peripheral receptors); (2) intermediate motor centers, e.g., pedal ganglia (control of ganglia within arms, the funnel, etc.), (3) higher motor centers, including supraesophageal lobes such as basal (movement of eyes and body), vertical, and superior frontal lobes and the brachial lobes (arm movements); and (4) receptor analyzers including the optic lobes, inferior frontal and posterior buccal lobes, statocyts, etc. The pallioviscerial lobes control the viscera and mantle (jet), the posterior chromophore lobes color-regulation by chromophores, and the superior buccal lobes the jaws. The superior and inferior frontal lobes appear to be important in visual learning and in touch learning, respectively.[1-3] The vertical lobe appears to be involved in both tactile and visual learning, and is important in short-term memory.

Much information concerning the functions of different regions of *Octopus* brain is the result of studies involving electrical stimulation and/or lesion or removal of specific brain areas.[1-3,15] For example, after removal of the vertical lobe of *O. vulgaris* the animal has difficulty in learning not to attack an object from which it has received a shock.[27,28] These studies suggest that the vertical lobe is involved in restraint and the integration of pain impulses into learned behavior. Removal of the subfrontal lobe, a structure containing a large number (about 6 million) of tiny amacrine cells and a much smaller number of larger neurons serving as output from the lobe, causes a major impairment of touch learning. The animal exhibits a discrimination failure and has an inability to learn to reject one object while continuing to take another.[29]

## B. *Mytilus edulis*

The general anatomical features of *M. edulis* are shown in Figure 4. Illustrated are the external anatomy (A), the internal anatomy after removal of shell (B), a view of the mantle

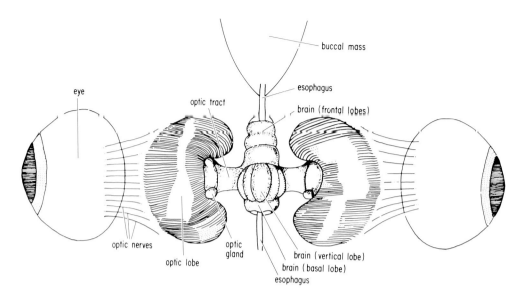

FIGURE 2A.   Diagram of the CNS of *Octopus bimaculatus* as seen from above. The relationships of the dorsal supraesophageal lobes of brain (frontal, vertical, basal), optic lobes, and eyes are shown. The eophagus passes through the brain, dividing the supra- from the subesophageal lobes (not shown). The optic lobes are connected to the brain (supraesophageal) by the optic tract. The optic gland is shown in relation to the optic tract and optic lobe. Next to each optic gland, also on the dorsal surface of the optic tract are two additional lobes (not shown): the peduncle lobe and the olfactory lobe. Surrounding the supra- and subesophageal lobes is a cranium, a box with lateral ridges composed of hard cartilagenous material, making an orbit in which lie the optic lobes and eyes. There is a small space between the cranium and the nervous tissue (the brain) containing a jelly-like connective tissue and small star-shaped cells. An extensive lobulated mass of soft tissue surrounding each optic called a white body, must be dissected away to clearly reveal the optic lobe and optic tract. The optic nerves arising from the retina emerge as compact bundles, each passing through a hole in the sclera, and travelling in an organized crossed arrangement to the surface of the optic lobe. There is neuronal input from optic lobe to retina as well as from retina to optic lobe.

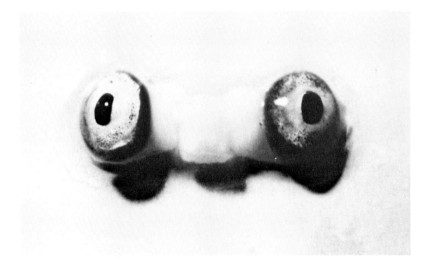

FIGURE 2B.   Photograph (from above) of dissected brain and optic lobes (with eyes attached). Orientation is about the same as in Figure 2A.

cavity after removal of the animal's right side (C), and a view of the circulatory and digestive systems (D). A detailed description of the gross neuroanatomy of *M. edulis* has been described by White.[30] For the purpose of this review, we will address only those elements of the nervous system that have been subsequently studied (Figure 5).

The "central" nervous system of *M. edulis* consists of 3 pair of ganglia, the cerebral, visceral,[14] and the pedal. The cerebral ganglia are triangular in shape, and their color can vary from milky white to a bright red-orange. These ganglia are connected by a cerebral commissure which passes over the animal's esophagus. An anterior pallial nerve which originates from the anterior angle of each ganglion innervates the labial palps, anterior adductor muscle, and mantle margin. Lateral to the anterior pallial nerve is the optic nerve which innervates an "eye" spot at the base of the first inner gill filament. A buccal nerve originates from the anterior end of the cerebral ganglion and terminates in the labial palps. Both the cerebropedal and cerebrovisceral connectives arise from the posterior apex of the cerebral ganglion as a common trunk. The connective passes posteriorly over the anterior byssus retractor muscle; upon reaching the muscle's lateral margin it separates into the two connectives. The cerebropedal connective passes posteriorly and ventrally to the anterior-most end of the pedal ganglion. The pedal ganglia lie at the dorsal medial aspect of the anterior byssus retractor muscles, just anterior to the base of the foot. The pedal ganglia are larger than the cerebral ganglia and are more rounded in appearance with the same type of coloration. The pedal commissure is extremely small. It may be difficult to see because of its size. The ganglia thus lie close together. Each ganglia gives off the following nerves: (1) the pedal nerve, which innervates the foot; originates from the ventral posterior surface; (2) the ventral byssus retractor nerve, which innervates the byssus organ and muscles; arises from the posterior ventral side of the ganglion; (3) the dorsal byssus retractor nerve, which also innervates the upper byssus muscles; arises from the posterior dorsal surface of the ganglion. From the point of the bifurcation of the cerebropedal and visceral connective, the latter connective passes posteriorly, giving off fine branches which go to innervate the digestive structures, genital organs, and kidney. The cerebrovisceral connective enters the visceral ganglia at its most anterior point. The visceral ganglia are the largest ganglia found in the organism. They appear to be "rounded" triangles with the same type of coloration previously described. They lie next to the posterior adductor muscle on its anterior ventral surface. The visceral commissure, which is relatively long and thick, gives rise to a posterior pedal nerve at its center. It is of interest to note that the distribution of the nerves which originate from the visceral ganglia is not always identical for each ganglion. Each ganglion gives off the following nerves: from the posterior surface arises the posterior pallial nerve which gives rise to three branches, (posterior ventral pallial nerve, posterior dorsal pallial nerve, and the siphon nerve). Both the ventral and dorsal pallial nerves fuse and then innervate the ventral surface of the mantle. The siphon nerve innervates the siphons and also fuses into another large nerve which emanates from the anterior medial portion of the visceral ganglion, the dorsal pallial nerve. The latter nerve sends fibers to innervate the viscera, including the pericardium and heart. The dorsal pallial nerve appears to merge with the posterior dorsal pallial nerve anatomically before going to innervate the viscera. The posterior renal nerve, which parallels the course of the cerebrovisceral connective, innervates the renal organs. The branchial nerve leaves the visceral ganglion laterally and runs to the gill axis where it gives off elements to each gill filament for innervation of the ciliated epithelium.[14,31,32] The posterior adductor nerve innervates the posterior adductor muscle, and arises from the visceral ganglion laterally.

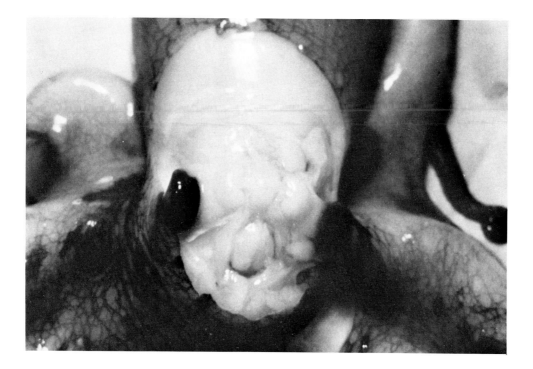

FIGURE 2C.    Photograph of exposed brain and optic lobes *in situ*. Dorsal cartilagenous material has been dissected away. Esophagus can be seen below and buccal mass above brain. Orientation is about the same as in Figure 2A.

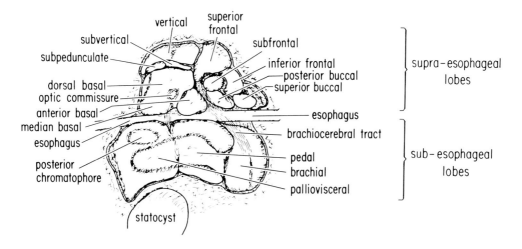

FIGURE 3.    Sagittal section (diagramatic representation of supra- and subesophageal lobes of *Octopus bimaculatus*. The anterior subesophageal masses (brachial lobes) on each side are connected by a commisure above the esophagus. This and other interconnections serve to integrate the sub- and supraesophageal regions of brain. Not shown here or in Figure 2 is a periesophageal lobe that runs along the sides of the brain.

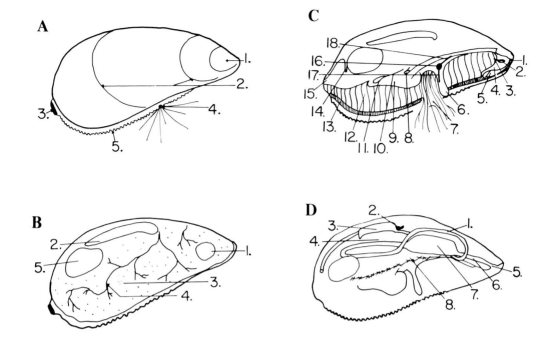

FIGURE 4.    Anatomy of *Mytilus edulis*. (A) External anatomy, (1) umbo, (2) growth lines, (3) exhalent siphon, (4) byssus threads, (5) mantle tentacles and area where water enters internal cavities; (B) removal of shell on right side, (1) anterior adductor muscle, (2) foot retractor muscle, (3) gonads, (4) mantle arteries, (5) posterior adductor muscle, (C) view of the mantle cavity following removal of the animal's right side, (1) mouth, (2) cerebral ganglion, (3) cerebiovisceral connective, (4) inner and (5) outer labial palps, (6) foot, (7) byssal threads, (8) urino-genital papilla, (9) visceral mass, (10) cerebrovisceral connective, (11) visceral ganglion, (12) inner gill and (13) outer gill lamella, (14) anus, (15) exhalent siphon, (16) pedal ganglion, (17) pedal nerve, (18) cerebropedal connective; (D) view of circulatory and digestive systems, (1) intestine, (2) aorta, (3) ventricle, (4) right atria, (5) mouth, (6) esophagus, (7) stomach, (8) kidney.

## IV. NEUROTRANSMITTER AND NEUROMODULATOR SYSTEMS

### A. Introduction

It is expected that if specific age-related changes in neuronal function occur, these changes will be reflected in and possibly due to changes in specific neurotransmitter or neuromodulator systems and/or in other neurochemical properties. It now is evident that most, if not all neurotransmitter and neuromodulator substances present in vertebrates are also present in the same or closely analogous form in invertebrates. Dopamine, for example, is present in planaria, trematodes, nematodes, annelids, molluscs, arthropods, echinoderms, and insects. This widespread occurrence of transmitter substances is the case not only for the classical amine and amino acid transmitters but also for neuropeptides. In addition, neuropeptides such as the molluscan cardioexcitatory peptide (FMRF amide), originally found in invertebrates may also be present in the vertebrate (including mammalian) CNS.[33]

### B. *Octopus*

The CNS of cephalopods contains many putative transmitter substances including dopamine (DA), norepinephrine (NE), 5-hydroxytryptamine (5-HT), octopamine, histamine, tyramine, taurine, acetylcholine (Ach), γ-aminobutyric acid (GABA), glutamate, aspartate,

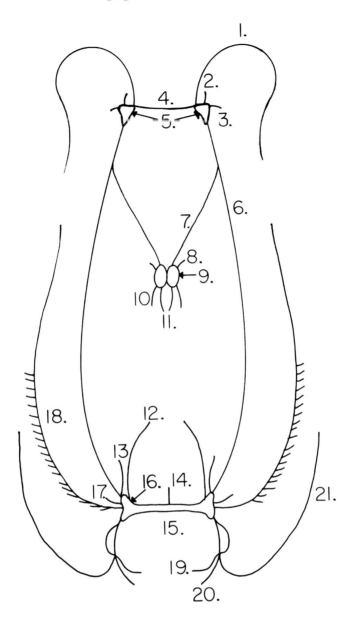

FIGURE 5.  Scheme of the nervous system of the marine mussel *Mytilus edulis*. (1) anterior pallial nerve, (2) buccal nerve, (3) optic nerve, (4) cerebral commissure, (5) cerebral ganglion, (6) cerebro-visceral connective, (7) cerebro-pedal connective, (8) dorsal byssus retractor nerve, (9) pedal ganglion (pedal commissure between ganglia), (10) ventral byssus retractor nerve, (11) pedal nerve, (12) dorsal pallial nerve, (13) posterior renal nerve, (14) posterior pedal nerve, (15) visceral commissure, (16) visceral ganglion, (17) posterior adductor nerve, (18) branchial nerve with fibers entering gill filaments, (19) posterior dorsal pallial nerve, (20) siphon nerve, (21) posterior ventral pallial nerve.

glycine and alanine.[34,35] The distribution of DA, NE, octopamine, and 5-HT in the CNS of several cephalopods including *O. vulgaris* has been studied.[35-37] Different distribution patterns were found for each of these amines. Very high concentrations, particularly of DA (up to 15 μg/g tissue) and NE (up to 10 μg/g tissue) were present in many lobes of brain and in optic lobes. Histofluorescence studies have confirmed the widespread occurrence of catecholamines in *Octopus* brain.[38] Amine concentrations in the optic lobe were decreased by severing the optic tract but not by cutting the optic nerves.[37] We have found DA and NE concentrations of 19.9 and 8.4 μg/g tissue respectively in optic lobe of *O. bimaculatus*.[39] Corresponding values were 2.4 and 0.6 μg/g for retina of *O. bimaculatus*. Thus, as is the case for mammalian retina, the *Octopus* retina contains appreciable amounts of DA but little or no NE. The concentrations of DA and NE in the *Octopus* brain and optic lobe are approximately the same as the concentration of DA in rat striatum and the concentration of DA in *Octopus* retina is at least as great as that in mammalian retina. Thus, the total amount of DA in the *Octopus* CNS actually exceeds that present in rat brain. Also, we found that in *O. bimaculatus* the optic lobe and (combined) supraesophageal lobes synthesized both DA and NE from $^{14}$C-tyrosine, whereas the retina synthesized only labeled DA in vitro. Retinas of *O. joubini* and *O. vulgaris* were previously shown to synthesize appreciable amounts of DA but very little NE.[40] Histofluorescence studies (unpublished) indicate that DA is localized in nerve endings in *Octopus* retina; the retinal DA probably arises from neurons with cell bodies located in the optic lobes. The optic lobes, both in structure and function, appear to be analogous, at least in part, to the inner (neural) retina of vertebrates.

Since the *Octopus* nervous system contains such high concentrations of DA, one would predict that DA receptors would also be present. We have reported that homogenates of retina of *O. bimaculatus* contains adenylate cyclase with high specificity for DA.[41] DA also stimulated cyclic AMP formation in intact pieces of *Octopus* retina incubated in vitro (Table 1). In the mammalian nervous system, DA receptors constitute heterogenous populations of sites. The strongest evidence for this heterogeneity is for a population of sites ($D_1$) mediating DA-stimulation of adenylate cyclase and for another population ($D_2$) either not linked or inhibitory to adenylate cyclase but most readily assessed by binding of antagonist (e.g., $^3$H-spiroperidol) or agonist (e.g., $^3$H-2-amino, 6,7 dihydroxy-1,2,3,4-tetrahydronaphthalene [$^3$H-ADTN] radioligands to specific sites.[42,43] $D_2$ Receptor sites in the mammalian CNS appear to exist in both agonist and antagonist conformations.[42] In *Octopus* brain (supraesophageal lobes), optic lobe (Figure 6), and retina (Figure 7) we have detected the agonist form of the $D_2$ receptor using the ligand $^3$H-ADTN (Figures 6 and 7) but thus far have not been able to detect the antagonist form of this receptor using $^3$H-spiroperidol as ligand.[44] As will be discussed later, since age-related changes in DA transmitter systems and DA receptors have been reported to occur in the mammalian brain, it is of some interest that DA and DA receptors are so prevalent in the *Octopus* nervous system.

Although the importance of aminergic systems in invertebrates has been evident for some time, the presence and role of peptide neurotransmitters and neuromodulators in invertebrates has only recently been considered.[45,46] Met-enkephalin-like immunoreactivity and FMRF amide (molluscan cardio excitatory peptide)-like immunoreactivity have been found to coexist in nerve terminals and neurosecretory granules of the vena cava of *O. vulgaris*.[47-49] Somatostatin-like immunoreactivity has been localized histochemically in the optic lobe and stellate ganglia of squid (*Loligo pealei*),[50] low levels of somatostatin-like immunoreactivity also appear to be present in the CNS of *O. bimaculatus*.[51] Recently, we have detected the presence of an adenylate cyclase stimulated by vasoactive intestinal peptide (VIP) as well as immunoreactive VIP (measured by radioimmunoassay) in the brain of *O. bimaculatus*.[52]

Indirect evidence for opioid receptors comes from studies of the influence of opioid agonists on the release of labeled dopamine from *O. bimaculatus* brain tissue slices incubated in vitro.[46] The opioid compounds morphine and D-ala$^2$-metenkephalin suppressed potassium-stimulated release from supraesophageal lobe tissue as well as from subdissected vertical, basal, and frontal lobes of brain. The inhibitory effects of the opioids were not seen in the presence of naloxone. Release of $^3$H-5-HT was not altered by opioids. The opioid effects were postulated to be directly mediated by presynaptically localized opiate receptors. Opioids have previously been shown by a number of investigators to inhibit release of DA and other transmitters in the mammalian central and peripheral nervous systems.[46]

**Table 1**
**DOPAMINE-STIMULATED CYCLIC AMP FORMATION**
**IN *OCTOPUS BIMACULATUS* RETINA**

| Additions | pmoles cyclic AMP/ mg protein | | |
|---|---|---|---|
| Control | $22.9 \pm 1.2$ | (22) | $\Delta$ |
| Dopamine | | | |
| $10^{-7}M$ | $28.5 \pm 1.0$ | (4) | 5.6 |
| $10^{-6}M$ | $32.4 \pm 0.8$ | (10) | 9.5 |
| $10^{-5}M$ | $44.4 \pm 1.6$ | (21) | 21.5 |
| $10^{-4}M$ | $42.1 \pm 2.5$ | (4) | 19.3 |
| Norepinephrine | | | |
| $10^{-5}M$ | $25.5 \pm 4.9$ | (4) | 3.6 |
| $10^{-4}M$ | 35.8 | (2) | 12.9 |
| Epinephrine, $10^{-5}M$ | $28.9 \pm 3.2$ | (4) | 6.0 |
| Lysergic acid diethylamide, $10^{-6}M$ | 23.8 | (2) | 0.9 |
| Dopamine, $10^{-5}M$ plus | | | |
| Pimozide | | | |
| $10^{-9}M$ | $20.1 \pm 3.0$ | (4) | 0 |
| $10^{-8}M$ | 19.4 | (2) | 0 |
| $10^{-7}M$ | 19.2 | (2) | 0 |
| Haloperidol | | | |
| $10^{-7}M$ | $32.4 \pm 1.1$ | (4) | 9.5 |
| $10^{-5}M$ | $24.4 \pm 1.2$ | (4) | 1.5 |
| Fluphenazine | | | |
| $10^{-7}M$ | $32.8 \pm 1.4$ | (4) | 9.9 |
| $10^{-5}M$ | $27.5 \pm 2.8$ | (4) | 3.1 |
| Lysergic acid diethylamide, $10^{-6}M$ | 23.1 | (2) | 0.2 |

*Note*: Retinal tissue was incubated for 30 min at 15° in artificial sea water (Instant Ocean) plus glucose. Values are means ± SEM (number of separate incubations in parentheses). Inactive at $10^{-5}M$ serotonin, N-$CH_3$-dopamine, isoproterenol, 2-chloradenosine.[44]

### C. *Mytilus edulis*

The ganglia of *M. edulis* contain the biogenic amines 5-HT, DA, and NE.[49] Considerable evidence suggests that all three amines function as neurotransmitters both centrally and peripherally in this organism.[14,31,32,53-56] Histologically, the ganglia of *M. edulis* follows the typical pattern for Pelecypods, having a cortex surrounding a central neuropile and containing no large cell bodies or fibers (Figure 8, A, B).[14,57] Utilizing the histochemical technique of Falck et al.,[58] The specific localization of various monoamines in the ganglia was demonstrated.[14] Of the fluorescent cells, 5-HT predominated in the cerebral ganglion and catecholamine predominated in the visceral and pedal ganglia (Figure 8, C,D).[14,55] There was a net flow of 5-HT in the cerebrovisceral connective from the cerebral to the visceral ganglion and a net flow of DA in the opposite direction. The visceral ganglion supplies the individual gill filaments with nerve fibers containing 5-HT and DA.[14,30] It also has demonstrated by histofluorescence that exogenously supplied DA and 5-HT could be accumulated by both types of cells.[14] More recently, the kinetics for uptake of tritiated 5-HT, DA, and NE has been studied.[59-61] For each of these amines a high-affinity uptake₁ component and lower-affinity uptake₂ component was demonstrated. The $Km_1$ and $Vmax_1$ values (pedal ganglia) for 5-HT were 0.2 $\mu M$ and 0.15 nmol/g/min whereas $Km_2$ and $Vmax_2$ values were 1.1 $\times$ $10^{-4}$ $M$ and 5.46 nmol/g/min, respectively. The same relationship, with slight differences in the actual values, was shown for DA and NE. 3-Chlorimipramine was a potent and select inhibitor of high-affinity 5-HT uptake, as was benztropine for DA uptake. Amine uptake

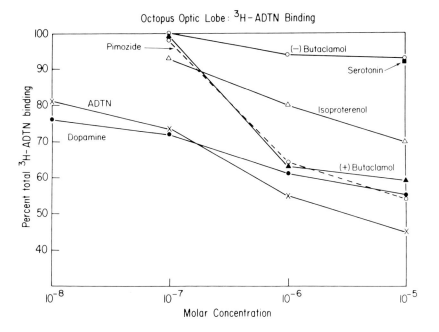

FIGURE 6. Competition by various drugs for ³H-ADTN binding sites in optic lobe of *Octopus bimaculatus*. Optic lobes were homogenized and membrane fractions assessed for binding of ³H-ADTN using a modification of procedures previously described for mammalian CNS.[42] In saturation studies carried out with increasing concentration of ³H-ADTN, Scatchard analysis indicated a $B_{max}$ value of 118 fmoles/mg protein and a $K_D$ value of 4.8 n$M$.[44]

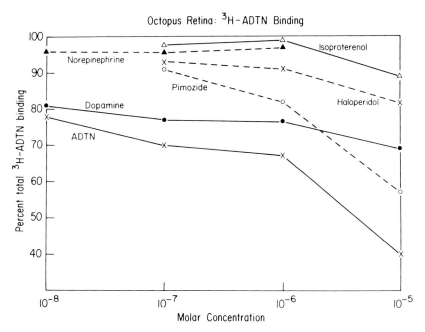

FIGURE 7. Competition by various drugs for ³H-ADTN binding sites in retina of *Octopus bimaculatus*.[44] Retinas were homogenized and the membrane fractions assessed for binding of ³H-ADTN using a modification of procedures previously described for mammalian CNS.[42]

was dependent upon the presence of sodium in the medium. In other molluscs, both high- and lower-affinity reuptake mechanisms have been demonstrated.[62-65] These mechanisms in molluscs are quite similar to those found in mammals.[66] The pedal ganglia of *M. edulis* also

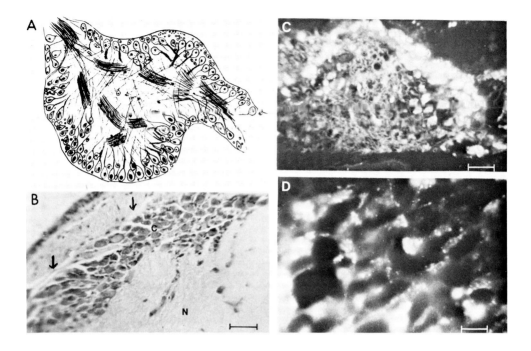

FIGURE 8.   (A) Ganglionic organization of *Mytilus edulis*. The three regions present (outside to center) are the connective tissue sheath, the cortex, and the central neuropil. There are large and small nerve cell bodies (width, 25-5 μm, respectively) in the cortex. Nerve tracts and synaptic contacts can be found within the neuropil with a few scattered nerve cell bodies. The nerve cells appear to be oriented with their processes coming off in the direction of the neuropil. Larger nerve cell bodies or fibers are not found in *M. edulis* ganglia. (B) Representative photomicrograph of a hematoxylin and eosin stained section through the cerebral ganglion showing the relationships of the connective tissue sheath (arrows), the cortex and the neuropil (N) (Scale bar = 20 μm). (C) Representative histofluorescent photomicrograph of a portion of the cerebral ganglion after 2 days exposure of whole animals to nialamide in the bathing medium. Cortical cells (bar = 20 μm) and fibers show enhanced fluorescent intensity of either yellow (serotonin) or green (catecholamines). (D) Representative histofluorescent photomicrograph of the pedal ganglia neuropil following treatment of the animal with 6-OHDA (6-hydroxydopamine; bar = 5 μm). The normally green-fluorescing neuropil was found to be devoid of the green fluorescing material and appeared to contain more yellow-fluorescing structures.[55] Both (C) and (D) above demonstrate that the ganglia of *M. edulis* are freely accessible to drugs and the various morphological effects of various agents can be readily visualized.

contain, in addition to DA itself and a system for uptake of DA, an adenylate cyclase that is stimulated by DA and antagonized by DA antagonists such as haloperidol and fluphenazine.[67]

Opioid peptides and alkaloids have been found to increase the concentration of DA in the pedal ganglia of *M. edulis*,[68] as well as in certain ganglia of the freshwater mollusc *Anodonta cygnea*,[69] and the land snail *Helix pomatia*.[70] The relative potencies of opioids in producing this effect and the blockade of the effect by the opioid antagonist naloxone strongly suggested the involvement of an opiate receptor mechanism. As was described earlier for brain tissue of *O. bimaculatus* incubated in vitro, the potassium-stimulated release of labeled DA from the pedal, cerebral, and visceral ganglia of *M. edulis* was inhibited by D-ala²-met-enkephalin and morphine, and this effect was also prevented by naloxone.[46] Ganglia of *M. edulis* were found to contain high-affinity, stereo-specific binding sites for opioid agonists and antagonists.[71,72] Opioid compounds such as etorphine and enkephalins inhibited DA-stimulated adenylate cyclase in the pedal ganglia of *M. edulis*,[67] indicating a colocalization of at least one class of opiate and DA receptors (D₁) linked to adenylate cyclase. Recently, enkephalin-like immunoreactivity was detected in proximity to DA-containing structures in the pedal ganglia of *M. edulis*.[73] Finally, peptides extracted from pedal ganglia and fractionated by high-pressure liquid chromatography (HPLC) had the same retention times as met- and leu-enkephalin as well as opioid activity in bioassay systems.[74]

## V. SEASONAL CHANGES

Biochemical and physiological processes found in living organisms appear to have a dynamic range of response potential. There appears to be incorporated into these systems a dynamic reactive capacity which is modifiable within certain pre-established parameters. As an organism progresses through its life cycle, the systems operate at different levels, however, all within the normal dynamic range or capacity. This dynamic reactive capacity may be observed when examining seasonal variations of various biochemical systems. In *M. edulis* ganglia, the monoamine levels vary depending on the season. Summer levels of 5-HT, DA, and NE were approximately 2-fold higher compared to winter months, in animals harvested from Long Island Sound.[75] Reuptake systems for 5-HT, DA, and NE have also been shown to vary seasonally.[76] The 5-HT high-affinity uptake component increased 5-fold in the winter when compared to spring, fall, and summer values. In July, the $Vmax_1$ value for 5-HT was 17 times greater than that determined during other times of the year. In addition, the potassium-stimulated release of various neurotransmitters varies seasonally.[77] In summer and winter months 50 m$M$ KCl is relatively ineffective in releasing 5-HT, DA, and NE from *M. edulis* neural tissues as compared to spring and fall months (N.Y.C. area).

The ability of opioid peptides to alter endogenous DA levels in the ganglia of *M. edulis* was found to be seasonal in nature.[78] In the spring and fall D-ala$^2$-met-enkephalin and met-enkephalin caused DA levels to increase while in the summer they did not. Enkephalin radioligand binding followed a similar pattern; that is, binding was highest in the spring and fall and lowest in the summer.[78] It is of interest to note that seasonal variations in opioid binding have been demonstrated in mammals.[79]

These results suggest that there is a rather broad set of overall biochemical as well as physiological parameters by which this organism survives. They also demonstrate that the reactive capacities of the various neural systems are extremely dynamic. These seasonal alterations in neuronal function may well be connected to the animal's reproductive cycle as well as other systems and/or cycles. For instance, the reproductive cycle consists of several phases; development of resting gonads (fall), gametogenesis (winter), spawning (spring), rapid gametogenesis (early summer), and then a resting state (for Long Island Sound, specifically Northport area).

The relatively short lifespan of *Octopus* is also intimately tied in with a variety of seasonal changes in the environment involving water temperature, food supply, etc. These external events determine spawning and the entire timing of the life cycle.[3] Biochemical and pharmacological evidence for seasonal variation has not yet been obtained for *Octopus* as it has for *M. edulis*. It is clear, however, that any investigation of age-related changes in *Octopus* as well as in the mussel must take into account the possibility of major seasonal variation in the parameters under investigation.

## VI. AGING STUDIES

### A. Advantages and Disadvantages of the Models

*Octopus* and *M. edulis* have nervous systems containing transmitters and related receptors that in the mammalian CNS exhibit age-related changes (see Section VI. C. below). The complex CNS of *Octopus* is particularly rich in aminergic transmitters and thus is an interesting model in general for investigation of the neurobiology and neuropharmacology of these transmitters. Perhaps the most serious disadvantage of the *Octopus* for studies of aging at present is the very complexity of the animal's nervous system and the many gaps that now exist in our understanding of its function. Another disadvantage is that it is often difficult to maintain the animals for long periods of time in the laboratory setting. This may be circumvented to an extent simply by obtaining animals from the wild at different ages (basing the age of mature animals on size, time of year, and place of capture) and holding them for defined periods of time for investigation. The choice of species of *Octopus* will depend on several factors including availability, whether the animals are to be maintained in an open or closed system, and whether the animals are to be hatched and reared for study throughout the lifespan, or to be studied primarily during the adult period of life. The relatively short and well-defined lifespan of *Octopus* is of definite advantage for studies of aging in this animal.

*M.edulis* has a less complex and therefore more readily analyzable nervous system. While the lifespan of *M. edulis* raised in a particular environment may be fairly well defined, the actual determinants and potential limits of lifespan of this animal are not well known. On the other hand, *M. edulis* is readily maintained in the laboratory or in the wild. Also, *M. edulis* has already been found to undergo age-related changes (reviewed in Section VI.B. below). It should be of considerable interest to further elucidate the basis and functional significance of these changes. Similar studies have not as yet been carried out in *Octopus*.

## B. Age-Related Changes in the Nervous System of *Mytilus Edulis*

The influence of age on several aspects of neuronal function and biochemistry in *M. edulis* are currently being investigated. Studies carried out thus far have included examination of biogenic amine uptake and levels, the influence of opioids on biogenic amine concentration, the concentration and affinity of opiate receptors, and the activity of DA-stimulated adenylate cyclase.

Ganglia from animals 4 years of age had significantly lower affinity and maximal accumulation values for DA and NE uptake than did ganglia from animals 1 and 2 years of age. These effects were selective for the high-affinity uptake component and neither high- nor low-affinity uptake of 5-HT was affected.[61] Representative data for DA are shown in Table 2. A decrease with age in the dependency of amine uptake on Na$^+$ was also noted. In addition, the inhibition of DA uptake by benztropine and the inhibition of NE uptake by desmethylimipramine declined with age. Concentrations of DA and NE in the ganglia, however, did not decline with age (Table 3).[61]

It will be necessary to measure turnover of amines directly in order to definitively assess whether or not amine metabolism is functionally altered in the older animals. However, another approach has provided indirect evidence at least that the functional regulation of amine metabolism is altered in older animals. As indicated earlier, topically applied opioid peptides increase DA levels in ganglia of *M. edulis*, probably by inhibiting the release of DA at the synapse.[46,68] In both 1- and 2-year-old animals, the concentration of DA in the visceral ganglia rose by over 40% when the ganglia were exposed to 30 μ*M* D-ala$^2$-met-enkephalinamide or 10 μ*M* etorphine. When the effect of these opioids was tested in 4-year-old animals, there was a significantly smaller increase, approximately 20%, in the concentration of DA (Table 3).[12] In addition, at the highest administered dose of either opioid, the DA concentration of the 1- and 2-year-old animals in the visceral ganglia was significantly higher than that of the older animals.

A possible explanation for the age-related decrease in opioid regulation of DA concentrations might be that there is a change in the opioid receptors mediating this regulation. The characteristics of opioid receptors were studied using both agonist ($^3$H-etorphine) and antagonist ($^3$H-naloxone) radioligands in ganglia of 1-, 2-, and 4-year-old animals.[11,12] Specific and saturable high-afinity binding was demonstrable for all age groups. Figure 9 shows the specific binding of $^3$H-etorphine to visceral ganglion membranes from young and old animals. When binding data were analyzed by Scatchard plot, a single class of high-

## Table 2
## KINETIC CONSTANTS FOR TRITIATED
## DOPAMINE ACCUMULATION BY PEDAL
## GANGLIA OBTAINED FROM *MYTILUS EDULIS*
## OF DIFFERENT AGES

| Age Years | High affinity system (uptake₁) | | Low affinity system uptake₂ | |
|---|---|---|---|---|
| | $Km_1$ ($\mu M$) | $Vmax_1$ (n mol/g/min) | $Km_2$ ($10^{-4}M$) | $Vmax_2$ (n mol/g/min) |
| 1 | 2.6 | 1.03 | 2.4 | 41.9 |
| 2 | 2.8 | 1.05 | 2.3 | 40.1 |
| 4 | 7.8 | 0.71 | 2.4 | 39.7 |

*Note:* The kinetic constants were determined by computer from linear regression line $y = mx + b$. The ganglia were incubated for 30 min with varying concentrations of $^3H$ dopamine (0.025 to 100 $\mu M$). The data was further analyzed graphically by Lineweaver-Burk plots and the Michaelis-Menten equation.[59]

## Table 3
## EFFECTS OF OPIOIDS ON DOPAMINE LEVELS IN *MYTILUS EDULIS* AT DIFFERENT AGES

| Treatment | N | Dopamine ($\mu$g/g at age) | | |
|---|---|---|---|---|
| | | 1 Year | 2 Year | 4 Year |
| Untreated | 4 | 25.6 ± 0.5 | 26.1 ± 0.6 | 25.9 ± 0.6 |
| Vehicle | 5 | 26.4 ± 0.8 | 26.6 ± 0.9 | 26.1 ± 0.9 |
| D-ala²-met-enkepha-linamide | | | | |
| 1 $\mu M$ | 8 | 28.7 ± 0.8 | 28.9 ± 0.7 | 25.2 ± 0.6 |
| 5 $\mu M$ | 8 | 31.5 ± 0.9[c] | 30.9 ± 0.9[c] | 27.3 ± 0.2 |
| 10 $\mu M$ | 8 | 33.0 ± 1.1[a] | 33.8 ± 1.0[a] | 29.1 ± 0.8 |
| 30 $\mu M$ | 8 | 34.7 ± 1.3[a] | 34.8 ± 1.2[a] | 29.9 ± 0.9[ca] |
| Etorphine | | | | |
| 0.1 $\mu M$ | 5 | 264. ± 0.9 | 26.1 ± 0.8 | 25.9 ± 1.0 |
| 1 $\mu M$ | 5 | 28.7 ± 1.2 | 29.1 ± 0.9 | 26.1 ± 1.1 |
| 5 $\mu M$ | 5 | 34.7 ± 1.1[b] | 35.1 ± 1.3[b] | 29.2 ± 1.6 |
| 10 $\mu M$ | 5 | 36.3 ± 1.3[a] | 36.7 ± 1.3[a] | 30.8 ± 1.2[ca] |

*Note:* Drugs, at the indicated concentrations were applied topically in 10-$\mu\ell$ aliquots to the visceral ganglia in vivo. The visceral ganglia of four animals (N = one pooling of four animals) excised 60 min after the last drug application were pooled and assayed spectrofluorometrically for dopamine. Values are means ± S.E.M. The data obtained from replicate assays of pooled tissues were analyzed statistically by a one-tailed student's $t$-test.[12] Compared to vehicle A) $p < 0.001$, B) $p < 0.005$, C) $p < 0.05$.

[a] Four year compared to 2 year at the same opioid dose.

affinity sites was found with $K_D$ values of 4.2, 4.0, and 2.2 n$M$ for the 1-, 2-, and 4-year-old animals, respectively. While these $K_D$ values were not significantly different from one another, the receptor density was significantly decreased in the older age group. The 1- and 2-year-old animals had $B_{max}$ values of 90 and 96 pmol/g of protein, respectively. The 4-year-old animals had a $B_{max}$ value of 66 pmol/g of protein, representing a significant (30%

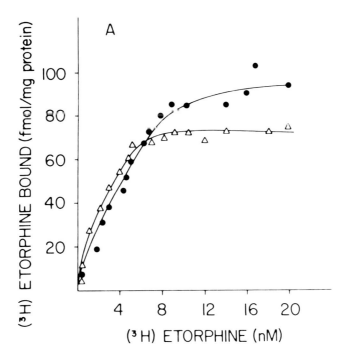

FIGURE 9.   Saturation analysis of specific ³H etorphine binding to *My-tilus edulis* visceral ganglia taken from animals at 2 (●) and 4 (Δ) years of age. Incubations were performed in the presence of 10 μ*M* dextrorphan or 10 μ*M* levorphanol for 90 min at 4°C. Each point represents the mean of three or four determinations. The concentration of protein and tissue wet weight for the various age groups was similar.[11]

decrease in the number of high-affinity sites. Binding of ³H-naloxone to visceral ganglion membranes also revealed a single class of high-affinity receptor sites for all age groups. Scatchard analysis revealed $K_D$ values of 8.1., 7.8, and 4.9 n*M* and $B_{max}$ values of 88, 84, and 65 pmol/g of protein for the 1-, 2-, and 4-year-old animals, respectively. Thus, again a significant (23%) decrease in the number of high-affinity binding sites was found in the older animals.

The loss of opioid receptors is apparently concomitant with the loss of DA responsiveness to opioid administration. These results also lend support to the possibility that opioid-receptor sites are present on dopaminergic neurons. The age-related decrease in opioid stimulation of DA levels involves both a decrease in the maximal responsiveness of the tissue and a reduced sensitivity to low concentrations of opioids. At present, the fraction of occupied opioid receptors necessary to increase the endogenous ganglionic DA content is not known.

The influence of aging on DA-stimulated adenylate cyclase activity of *M. edulis* ganglia has also been examined.[13] DA at 10 μ*M* stimulated adenylate cyclase activity to the same extent in pedal ganglion homogenates derived from 1- and 2-year-old animals, whereas the increase in activity due to DA was approximately 43% less in homogenates from 4-year-old animals (Figure 10). A loss of sensitivity of this system was also found for the inhibition of DA-stimulated adenylate cyclase by chlorpromazine, a DA-antagonist. In 1- and 2-year-old animals, 10 μ*M* chlorpromazine decreased DA-stimulated activity by about 40% whereas in older animals there was only a 10% decrease. Age differences in basal adenylate cyclase activity and in the stimulatory effect of the nonhydrolyzable GTP analogue, Gpp(NH)p, were not found.

The various age-related changes found to occur in *M. edulis* are summarized in Table 4. Since changes in the parameters studied did not occur between 1 and 2 years of age, the

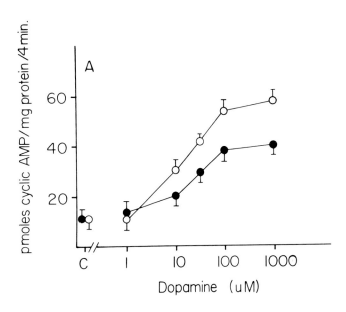

FIGURE 10.   The in vitro stimulation of adenylate cyclase by dopamine in pedal ganglia homogenates derived from *Mytilus edulis* 1 to 2 years old (○) and 4 years old (●). Control level of cyclic AMP formation for all age groups was 15.1 ± 1.6 pmol cyclic AMP/mg protein/4 min ± S.E.M.[13]

## Table 4
## SUMMARY OF NEURAL PARAMETERS EXAMINED THUS FAR IN REFERENCE TO AGING IN *MYTILUS EDULIS*

| Parameter | Changes during aging |
|---|---|
| Intraganglionic basal cyclic AMP levels | No change |
| Dopamine-stimulated adenylate cyclase | |
|   Maximal response to dopamine | Decrease |
|   Sensitivity to dopamine | Decrease |
|   Sensitivity to chlorpromazine | No change |
|   GTP-stimulated adenylate cyclase | No change |
| Opiate Mechanisms | |
|   Etorphine receptor affinity ($K_d$) | No change |
|   Naloxone receptor affinity ($K_d$) | No change |
|   Etorphine receptor density (B max) | Decrease |
|   Naloxone receptor density (B max) | Decrease |
|   Etorphine ability to stimulate dopamine levels | Decrease |

### Monoamine Accumulation

| | Serotonin | Dopamine | Norepinephrine |
|---|---|---|---|
| High-affinity component | | | |
|   Affinity | No change | Decrease | Decrease |
|   Amount accumulated | No change | Decrease | Decrease |
| Lower-affinity component | | | |
|   Affinity | No change | No change | No change |
|   Amount accumulated | No change | No change | No change |
| Sensitivity to omission of $Na^+$ from medium | No change | Decrease | Decrease |
| Sensitivity to omission of $K^+$ from medium | No change | Decrease | Decrease |

changes that took place in the older animals very likely represent postmaturational events rather than the continuation of a previously started decay process. One of the most predominant features or characteristics of aging in animals may be considered to be the alteration or weakening of intrinsic adaptive reflexes, thus reducing the organism's reactive capacity. Certainly the alterations or impairments demonstrated in *M. edulis* neural tissues involve inter- as well as intra-cellular communication mechanisms. This may in turn predispose an organism to disease or it may succumb to various stressful environmental factors. In conclusion, there appears to be a decrease of reliability of homeostatic regulation with age.

## C. Comparison with Vertebrate Systems: Possible Similarities and Differences

Studies of aging in general have been concerned with determinants of lifespan and cause of death, decreases in proliferative capacity of different cell types, and functional impairments associated with old age. The nature and extent of changes that occur with increasing age may differ considerably in different species and also in the various organs or tissues of a given species. Functionally important age-related changes may or may not also be important causes of death of the animal. In addition, a particular age-related pathological process may be truly unique to one species or to a group of species of animals. On the other hand, important features of senescence may be shared by different animals and there may in fact be aspects of aging that are common to vertebrates and inverebrates. In the human population there is evidence for age-related decrements in CNS function, including behavioral impairment and altered response to CNS drugs.[80-86] Certain of these decrements may apply to the general human population and be due to selective age-related alterations in neurotransmitter systems that also occur with aging in other mammalian species. However, there may be no close counterpart in other species to senile dementia of the Alzheimer's type, a human disease associated with, and possibly caused by loss of certain central cholinergic neurons.[86-89]

Age-related deficits of nigrostriatal, hypothalamic, and cortical dopaminergic function occur in several mammalian species. Decreases in DA levels, uptake and turnover, tyrosine hydroxylase activity, DA-stimulated adenylate cyclase, and radioligand binding to DA receptors have been reported.[81-85,90-92] It should be noted, however, that significant strain differences have been reported for nigrostriatal dopaminergic regulation in mice,[93] and that strain as well as species differences very likely exist in the extent of nigrostriatal aging in rodents.[84] While a decline in DA-stimulated adenylate cyclase in the brain may occur relatively late in the lifespan of the rabbit,[90] in different strains of rat, changes in this parameter either do not occur or occur only relatively early in life (before 12 months of age).[84]

Age-related changes in DA content and turnover in mammalian brain are generally small but may be of functional significance, particularly in regard to hypothalamic regulation. One of the more striking and consistent findings in experimental animals (rabbit, rat, mouse) is the decrease in striatal DA-receptor binding with age, a decrease that is more marked for the agonist radioligand, $^3$H-ADTN, than for the antagonist radioligand, $^3$H-spiroperidol.[84,91,92] This decrease does not appear to be due to neuronal cell loss or to alteration of DA metabolism presynaptically. A number of findings, including the relatively greater loss of guanine nucleotide-sensitive $^3$H-ADTN binding sites with age, suggest that aging results in a selective loss of a high-agonist-affinity form of $D_2$ receptor that is sensitive to guanine nucleotide and to sodium ion.[43,84,94] It is of interest that this form of $D_2$ receptor is also the first to appear during development in the rat striatum.[95] In addition, this form of $D_2$ receptor is selectively enhanced following destruction of the nigrostriatal DA pathway.[94] In normal-aged human brain there may also be a selective loss of $D_2$ agonist binding sites as assessed by $^3$H-ADTN binding.[83]

It should be evident from the above discussion that the age-related changes in the dopaminergic system in *M. edulis*, as summarized in Table 4, are similar in many respects to those changes found in mammalian brain.

Apart from considerations of aging, the properties of both aminergic and opioid transmitter systems in a broad sense appear to be similar in vertebrates and invertebrates. More detailed examinations seem likely to reveal both similarities and differences. Thus, there are significant pharmacological differences between DA-stimulated adenylate cyclase ($D_1$ receptor) in the mammalian CNS (including retina) and in *O. bimaculatus* retina.[44] The characteristics of radioligand binding to dopaminergic $D_2$ receptors is also different in mammalian and in *O. bimaculatus* brain. Whether or not there are in fact separate classes of $D_1$ and $D_2$ receptors in the *Octopus* CNS has not yet been established.

Opioid agonist binding in *M. edulis* ganglia, as in vertebrate neuronal tissue, is inhibited by $Na^+$ and this inhibition is reversed by $Mn^{++}$[71]. In *M. edulis*, visceral ganglia opioid agonists and antagonists bind to a single class of high-affinity ($K_D = 1 - 3$ n$M$) sites. However, in pedal ganglia both this class and a class of lower-affinity ($K_D = 6 - 11$ n$M$) sites are present. This latter class of sites exhibits a marked positive cooperativity of binding;[72] whether or not there are analogous sites in the mammalian nervous system is not certain.[72] The inhibitory effect of opioids on DA-stimulated adenylate cyclase of *M. edulis* ganglia does appear to resemble the interaction of opioids with DA-stimulated adenylate cyclase of monkey brain.[96,97]

Age-related decreases in opioid binding and opioid regulation of DA levels in *M. edulis* have been described earlier. It would be of interest to ascertain also whether or not there are age-related changes in the interaction of opioids with DA-stimulated adenylate cyclase in *M. edulis*. Decreased opioid receptor densities have been reported to occur in several regions of rat brain.[98] In contrast to that study, however, we have not found a significant change with age in opioid-receptor density in rat striatum.[99]

Further studies will be required to ascertain to what extent age-related changes in transmitters and related receptors in *M. edulis* ganglia and the mammalian CNS are similar. Given the great differences in neuronal organization and function between *M. edulis* and the mammalian species investigated, certainly one would expect there to be different patterns in the levels and function of specific transmitter systems during the lifetime of the animals. It is of significance, however, that there are great similarities in basic cellular mechanisms and in certain specific transmitter interrelationshps (e.g., for DA and opioids) in the invertbrate models and the mammalian systems.

## VII. CONCLUSIONS

The study of nervous systems of marine invertebrates such as *M. edulis* and *Octopus* greatly enhances our knowledge of the functioning of biological organisms so different from mammalian organisms, including man. Such investigations are important for our understanding of biological continuity, diversity, and evolution. They may reveal surprising similarities in the integrative processes regulating function and the life cycle in mammals and invertebrates. Also, studies of neuronal aging in the proposed invertebrate models should serve to focus attention on those aspects of neuronal processes and adaptive mechanisms necessary for survival and propagation of the organism. For these reasons, such studies have a general relevance to our understanding of life cycle events and the aging process in all animal species.

## ACKNOWLEDGMENTS

This work was supported by USPHS Grants NS-09649, AG-00374, AG-01400 (to M.H.M.), and by MH-17138 and MBRS Grant RR-08180 (to G.B.S.).

# REFERENCES

1. **Young, J. Z.,** *The Anatomy of the Nervous System of Octopus Vulgaris,* Oxford, 1971.
2. **Young, J. Z.,** *What Squids and Octopuses Tell Us About Brains and Memories,* American Museum of Natural History, N.Y., 1977.
3. **Wells, M. J.,** *Octopus: Physiology and Behavior of an Advanced Invertebrate,* Chapman and Hall, London, 1978.
4. **Voss, G. L.,** Present status and new trends in cephalopod systematics, *Symp Zool Soc. London,* 38, 49, 1977.
5. **Boletzky, S. v. and Boletzky, M. v.,** First results in rearing *Octopus joubine* Robson, *Verh. Naturforsch. Ges.,* Basel, 80, 56, 1969.
6. **Forsythe, J. W. and Hanlon, R. T.,** A closed marine culture system for rearing *Octopus joubini* and other large-egged benthic octopods, *Lab. Anim.,* 14, 137, 1980.
7. **Hanlon, R. T.,** Laboratory rearing of the Atlantic reef octopus, *Octopus briareus* Robson, and its potential for mariculture, *Proc. World Mariculture Soc.,* 8, 471, 1977.
8. **Hanlon, R. T., Hixon, R. F., and Forsythe, J. W.,** The 'Maciotritopus problem' solved: *Octopus defilippi* reared from a wild-caught, pelagic Macrotitopus, *Bull. Am. Mala. Union,* (Abstr.), 1980.
9. **Wodinsky, J.,** Hormonal inhibition of feeding and death in *Octopus:* control by optic gland secretion, *Science,* 198, 948, 1977.
10. **Seed, R.,** Shell growth and form in the *Bivalvia,* in *Skeletal Growth in Aquatic Organisms,* Rhoads, D. C. and Lutz, R. A., Eds., Plenum Press, N.Y., 1980.
11. **Stefano, G. B.,** Decrease in the number of high affinity opiate binding sites during the aging process in *Mytilus edulis (Bivalvia), Cell Mol. Neurobiol.,* 1, 343, 1981.
12. **Stefano, G. B.,** Aging: variations in opiate binding characteristics and dopamine responsiveness in subtidal and intertidal *Mytilus edulis* visceral ganglia, *Comp. Biochem. Physiol.,* 72C, 349, 1982.
13. **Stefano, G. B., Stanec, A., and Catapane, E. J.,** Aging: decline of dopamine-stimulated adenylate cyclase activity in *M. edulis (Bivalvia), Cell. Mol. Neurobiol.,* 2, 249, 1982.
14. **Stefano, G. B. and Aiello, E.,** Histofluorescent localization of serotonin and dopamine in the nervous system and gill of *Mytilus edulis (Bivalvia), Biol. Bull.,* 148, 141, 1975.
15. **Wells, M. J.,** Evolution and associative learning, in *Simple Nervous Systems,* Usherwood, P. N. R. and Newth, D. R., Eds., Edward Arnold, London, 1975, 446.
16. **Young, J. Z.,** The number and sizes of nerve cells in *Octopus, Proc. Zool. Soc. London,* 140, 229, 1963.
17. **Young, J. Z.,** The central nervous system of *Loligo.* I. The optic tract, *Philos. Trans. R. Soc. London Ser. B,* 267, 263, 1974.
18. **Young, J. Z.,** The nervous system of *Loligo.* II. Subaesophageal centres, *Philos. Trans. R. Soc. London Ser. B,* 274, 101, 1976.
19. **Young, J. Z.,** The nervous system of *Loligo.* III. Higher motor centres. The basal supraesophageal lobes, *Philos. Trans. R. Soc. London Ser. B,* 276, 351, 1977.
20. **Young, J. Z.,** The nervous system of *Loligo.* V. The vertical lobe complex, *Philos. Trans. R. Soc. London Ser. B,* 285, 311, 1979.
21. **Monsell, E. M. and Cottee, L. J.,** Retrograde intraaxonal transport of HRP by neurons in *Octopus, Brain Res.,* 181, 251, 1980.
22. **Monsell, E. M.,** Cobalt and HRP tracer studies in the stellate ganglion of *Octopus, Brain Res.,* 184, 1, 1980.
23. **Saidel, W. M.,** Evidence for visual mapping in the peduncle lobe of *Octopus, Neurosci. Lett.,* 24, 7, 1981.
24. **Saidel, W. M.,** Relationship between photoreceptor terminations and centrifugal neurons in the optic lobe of *Octopus, Cell Tissue Res.,* 204, 462, 1979.
25. **Saidel, W. M.,** Connections of the *Octopus* optic lobe: an HRP study, *J. Comp. Neurol.,* 206, 346, 1982.
26. **Colmers, W. F.,** The central afferent and efferent organization of the gravity receptor system of the stratocyst of *Octopus vulgaris, Neuroscience,* 7, 461, 1982.
27. **Boycott, B. B. and Young, J. Z.,** A memory system in *Octopus vulgaris* Lamarck., *Proc. R. Soc. London Ser. B,* 143, 449, 1955.
28. **Wells, M. J. and Wells, J.,** The effect of lesions to the vertical and optic lobes on tactile discrimination in *Octopus, J. Exp. Biol.,* 34, 378, 1957.
29. **Wells, M. J. and Young, J. Z.,** The subfrontal lobe and touch learning in the octopus, *Brain Res.,* 92, 103, 1975.
30. **White, K. M.,** On typical British marine plants and animals. XXXI. *Mytilus.* Liverpool, *Mar. Biol. Comm. Mem.,* 31, 1, 1937.
31. **Catapane, E. J., Stefano, G. B., and Aiello, E.,** Pharmacological study of the reciprocal dual innervation of the lateral ciliated gill epithelium by the CNS of *Mytilus edulis, J. Exp. Biol.,* 74, 101, 1978.
32. **Catapane, E. J., Stefano, G. B., and Aiello, E.,** Neurophysiological correlates of dopaminergic cilio-inhibitory mechanism, *J. Exp. Biol.,* 8, 315, 1979.

33. **Weber, E., Evans, C. J., Samuelsson, S. J., and Barchas, J. D.,** Novel peptide neuronal system in rat brain and pituitary, *Science*, 214, 1248, 1981.
34. **Tansey, E. M.,** Neurotransmitters in the cephalopod brain, *Comp. Biochem. Physiol.*, 64C, 173, 1979.
35. **Barlow, J. J.,** Comparative biochemistry of the central nervous system, *Symp. Zool. Soc. London*, 38, 325, 1977.
36. **Juorio, A. V.,** Catecholamines and 5-hydroxytryptamine in nervous tissue of cephalopods, *J. Physiol. (London)*, 215, 213, 1971.
37. **Juorio, A. V. and Molinoff, P. B.,** The normal occurrence of octopamine in neural tissue of the *Octopus* and other cephalopods, *J. Neurochem.*, 22, 271, 1974.
38. **Matus, A. I.,** Histochemical localization of biogenic monoamines in the cephalic ganglia of *Octopus vulgaris*, *Tissue Cell*, 5, 591, 1973.
39. **Shapless, N. and Makman, M. H.,** Unpublished studies.
40. **Lam, D. M. K., Wiesel, T. N., and Kaneko, A.,** Neurotransmitter synthesis in cephalopod retina, *Brain Res.*, 82, 365, 1974.
41. **Makman, M. H., Brown, J. H., and Mishra, R. K.,** Cyclic AMP in retina and caudate nucleus: influence of dopamine and other agents, *Adv. Cyclic Nucleotide Res.*, 5, 661, 1975.
42. **Kebabian, J. W. and Calne, D. B.,** Multiple receptors for dopamine, *Nature* (London), 277, 93, 1979.
43. **Makman, M. H., Dvorkin, B., and Klein, P. N.,** Sodium ion modulates $D_2$ receptor characteristics of dopamine agonist and antagonist binding sites in striatum and retina, *Proc. Natl. Acad. Sci. U.S.A.*, 79, 4212, 1982.
44. **Makman, M. H. and Dvorkin, B.,** Unpublished studies.
45. **Leake, L. D. and Walter, R. J.,** *Invertebrate Neuropharmacology*, Halsted Press (John Wiley), N.Y., 1980.
46. **Stefano, G. B., Hall, B., Makman, M. H., and Dvorkin, B.,** Opioid inhibition of dopamine release from nervous tissue of *Mytilus edulis* and *Octopus bimaculatus*, *Science*, 213, 928, 1981.
47. **Martin, R., Frösch, D., Weber, E., and Voigt, K. H.,** Met-enkephalin-like immunoreactivity in a cephalopod neurohemal organ, *Neurosci. Lett.*, 15, 253, 1979.
48. **Voigt, K. H., Krehling, C., Frösch, D., Schiebe, M., and Martin, R.,** Enkephalin-related peptides: direct actions on the octopus heart, *Neurosci. Lett.*, 27, 25, 1981.
49. **Martin, R., Frösch, D., Kiehling, C., and Voigt, K. H.,** Molluscan neuropeptide-like and enkephalin-like material coexists in octopus nerves, *Neuropeptides*, 2, 41, 1981.
50. **Feldman, S. C.,** Immunohistochemical localization of somatostatin and calcium binding protein in squid and aplysia neurons, *Fed. Proc.* 41 (Abstr.), 8683, 1982.
51. **Feldman, S. C. and Makman, M. H.,** Unpublished studies.
52. **Sharpless, N. S., Longshore, M., and Makman, M. H.,** Unpublished studies.
53. **Stefano, G. B. and Catapane, E. J.,** Norepinephrine: its presence in the CNS of the bivalve mollusc *Mytilus edulis.*, *J. Exp. Zool.* 214, 209, 1980.
54. **Aiello, E. and Guideri, G.,** Relationships between serotonin and nerve stimulation of ciliary activity, *J. Pharmac. Exp. Ther.*, 154, 351, 1966.
55. **Stefano, G. B., Catapane, E. J., and Aiello, E.,** Dopaminergic agents: influence serotonin in molluscan nervous system, *Science*, 194, 539, 1976.
56. **Stefano, G. B., Hiripi, L., and Catapane, E. J.,** The effect of short and long term temperature stress on serotonin, dopamine and norepinephrine metabolism in molluscan ganglia, *J. Therm. Biol.*, 3, 79, 1978.
57. **Rawitz, B.,** Das Zentrale Nervensystem der Acephalin, *Jenai. Z. Med. Naturwiss.*, 20, 384, 1887.
58. **Falck, B., Hillarp, N., Thieme, S., and Torp, A.,** Fluorescence of catecholamines and related compounds condensed with formaldehyde, *J. Histochem. Cytochem.*, 19, 348, 1962.
59. **Burrell, D. E. and Stefano, G. B.,** Analysis of monoamine accumulations in the neuronal tissues of *Mytilus edulis (Bivalvia)*. I. Ganglionic variations, *Comp. Biochem. Physiol.*, 70C, 71, 1981.
60. **Brown, M., Burrell, D. E., and Stefano, G. B.,** Analysis of monoamine accumulation in the neuronal tissue of *Mytilus edulis (Bivalvia)* II. Pharmacological alteration of pedal ganglia monoamine uptake, *Comp. Biochem. Physiol.*, 70C, 215, 1981.
61. **Burrell, D. E. and Stefano, G. B.,** Analysis of monoamine accumulation of the neuronal tissues of *Mytilus edulis (Bivalvia)*. IV. Variations due to age, *Comp. Biochem. Physiol.*, 74C, 59, 1983.
62. **Ascher, P., Glowinski, J., Tauc, L., and Taxi, J.,** Discussion of stimulation-induced released of serotonin, *Adv. Pharmac.*, 6A, 351, 1968.
63. **Carpenter, D., Breese, G., Shanberg, S., and Kopin, I.,** Serotonin and dopamine distributions and accumulation in *Aplysia* nervous and non-nervous tissue, *Int. J. Neurosci.*, 2, 49, 1971.
64. **Hiripi, L., Rakonczay, Z., and Nemesok, J.,** The uptake kinetics of serotonin, dopamine and noradrenaline in pedal ganglia of the fresh-water mussel (*Anodonta cygnea* L. Pelecypode), *Ann. Biol.*, 42, 21, 1975.
65. **Osborne, N. N., Hiripi, L., and Neuhoff, R.,** The *in vitro* uptake of biogenic amines by snail (*Helix Pomatia*) nervous tissue, *Biochem. Pharmac.*, 24, 2141, 1975.

**188** *Invertebrate Models in Aging Research*

66. **Iversen, L. L.,** Uptake process for biogenic amines, in *Handbook of Psychopharmacology*, Vol. 3, Iversen, L. L., Iversen, S. D., and Snyder, S. H., Eds., Plenum Press, N.Y., 1975, 381.

67. **Stefano, G. B., Catapane, E. J., and Kream, R. M.,** Characterization of the dopamine stimulated adenylate cyclase in the pedal ganglia of *Mytilus edulis*: interactions with etorphine B-endorphin, DALA, and methionine enkephalin, *Cell. Mol. Neurobiol.*, 1, 57, 1981.

68. **Stefano, G. B. and Catapane, E. J.,** Enkephalins increase dopamine levels in the CNS of a marine mollusc, *Life Sci.*, 24, 1917, 1979.

69. **Stefano, G. B. and Hiripi, L.,** Methionine enkephalin and Morphine alter monoamine and cycle nucleotide levels in cerebral ganglia of the freshwater bivalvia *Anodonta cygnea*, *Life Sci.*, 25, 291, 1979.

70. **Osborne, N. N. and Neuhoff, V.,** Are there opiate receptors in the invertebrates?, *Pharm. Pharmacol.*, 31, 481, 1979.

71. **Stefano, G. B., Kream, R. M., and Zukin, R. S.,** Demonstration of stereo specific opiate binding in the nerve tissue of the marine mollusc *Mytilus edulis*, *Brain Res.*, 180, 440, 1980.

72. **Kream, R. M., Zukin, R. S., and Stefano, G. B.,** Demonstration of two classes of opiate binding sites in the nervous tissue of the marine mollusc *Mytilus edulis*, *J. Biol. Chem.*, 255, 9218, 1980.

73. **Stefano, G. B. and Martin, R.,** Enkephalin-like immunoreactivity in the pedal ganglion of *Mytilus edulis (Bivalvia)* and its proximity to dopamine containing structures, *Cell Tissue Res.*, 230, 167, 1983.

74. **Stefano, G. B. and Leung, M.,** Purification of opioid peptides from molluscan ganglia, *Cell. Mol. Neurobiol.*, 2, 347, 1982.

75. **Stefano, G. B. and Catapane, E. J.,** Seasonal monoamine changes in the central nervous system of *Mytilus edulis (Bivalvia)*, *Experientia*, 33, 1341, 1977.

76. **Hiripi, L., Burrell, D. E., Brown, M., Assanah, P., Stanec, A., and Stefano, G. B.,** Analysis of monoamine accumulation in the neuronal tissues of *Mytilus edulis (Bivalvia)*. III. Temperature and seasonal influences, *Comp. Biochem. Physiol.*, 71C, 209, 1982.

77. **Stefano, G. B.,** Unpublished studies.

78. **Stefano, G. B., Kream, R. M., Zukin, R S., and Catapane, E. J.,** Seasonal variation of stereospecific enkephalin binding and pharmacological activity in marine mollusc nervous tissue, in *Advances Physiological Sciences*, Vol. 22, Rozsa, K. S., Ed., Pergamon Press, London, 1980, 453.

79. **Codd, E. E. and Byrne, W. L.,** Seasonal variation in the apparent number of $^3$H-naloxone binding sites, in *Endogenous and Exogenous Opiate Agonists and Antagonists*, Way, E. L., Ed., Pergamon Press, N.Y., 1980, 63.

80. **Makman, M. H., Ahn, H. S., Thal, L. J., Dvorkin, B., Horowitz, S. G., Sharpless, N. S., and Rosenfeld, M.,** Biogenic amine-stimulated adenylate cyclase and spiroperidol-binding sites in rabbit brain: evidence for selective loss of receptors with aging, in *Parkinson's Disease — II: Aging and Neuroendocrine Relationships (Adv. Exp. Med. Biol.)*, Vol. 113, Finch, C. E., Potte, D. E., and Kenny, A. D., Eds., Plenum Press, N.Y., 1978, 211.

81. **Makman, M. H., Gardner, E. L., Thal, L. J., Hirschhorn, I. D., Seeger, T. F., and Bhargava, G.,** Central monoamine receptor systems: influence of aging, lesion and drug treatment, in *Neural Regulatory Mechanisms During Aging (Modern Aging Research)*, Vol. 1, Adelman, R. C., Roberts, J., Baker, G. T., Baskin, S. I., and Cristofalo, V. J., Eds., Alan R. Liss, N.Y., 1980, 91.

82. **Finch, C. E., Marshall, J. F., and Randall, P. K.,** Aging and basal ganglion functions, *Ann. Rev. Gerontol. Geriatr.*, 2, 49, 1981.

83. **Severson, J. A., Marcusson, J., Winblad, B., and Finch, C. E.,** Age-correlated loss of dopaminergic binding sites in human basal ganglia, *J. Neurochem.*, 39, 1623, 1982.

84. **Hirschhorn, I. D., Makman, M. H., and Sharpless, N. S.,** Dopamine receptor sensitivity following nigrostriatal lesion in the aged rat, *Brain Res.*, 234, 357, 1982.

85. **Joseph, J. A., Filburn, C., Tzankov, S. P., Thompson, J., and Engel, B. T.,** Age-related neustriatal alterations in the rat: failure of L-DOPA to alter behavior, *Neurobiol. Aging*, 1, 119, 1980.

86. **Bartus, R. T., Dean, R. L., Beer, B., and Lippa, A. S.,** The cholinergic hypothesis of geriatric memory dysfunction, *Science*, 217, 408, 1982.

87. **Davies, P. and Maloney, A. J. F.,** Selective loss of central cholinergic neurons in Alzheimer's disease, *Lancet*, 2, 1403, 1976.

88. **Davies, P. and Terry, R. D.,** Cortical somatostatin-like immunoreactivity in cases of Alzheimer's disease and senile dementia of the Alzheimer type, *Neurobiol. Aging*, 2, 9, 1981.

89. **Whitehouse, P. J., Price, D. L., Struble, R. G., Clark, A. W., Coyle, J. T., and DeLong, M. R.,** Alzheimer's disease and senile dementia, loss of neurons in the basal forebrain, *Science*, 215, 1237, 1982.

90. **Makman, M. H., Ahn, H. S., Thal, L. J., Sharpless, N., Dvorkin, B., Horowitz, S. G., and Rosenfeld, M.,** Evidence for selective loss of brain dopamine- and histamine-stimulated adenylate cyclase activities in rabbits with aging, *Brain Res.*, 192, 177, 1980.

91. **Thal, L. J., Horowitz, S. G., Dvorkin, B., and Makman, M. H.,** Evidence for loss of brain $^3$H-spiroperidol and $^3$H-ADTN binding sites in rabbit brain with aging, *Brain Res.*, 192, 185, 1980.

92. **Severson, J. A., and Finch, C. E.,** Reduced dopaminergic binding during aging in the rodent striatum, *Brain Res.,* 192, 147, 1980.
93. **Severson, J. A., Randall, P.K., and Finch, C. E.,** Genotypic influence on striatal dopaminergic regulation in mice, *Brain Res.,* 210, 201, 1981.
94. **Hirschhorn, I. D. and Makman, M. H.,** Altered characteristics of strital [$^3$H]ADTN binding following substantial nigra lesions, *Eur. J. Pharmacol.,* 83, 61, 1982.
95. **O'Connell, M. E. and Makman, M. H.,** Unpublished studies.
96. **Walczak, S. A., Wilkening, D., and Makman, M. H.,** Interaction of morphine, etorphine and enkephalins with dopamine-stimulated adenylate cyclase of monkey amygdala, *Brain Res.,* 160, 105, 1979.
97. **Walczak, S. A., Makman, M. H., and Gardner, E. L.,** Acetymethadol metabolities influence opiate receptors and adenylate cyclase in amygdala, *Eur. J. Pharmacol.,* 72, 343, 1981.
98. **Messing, R. B., Vasquez, B. J., Samaniego, B., Jensen, R. E., Martinez, J. L., and McGough, J. L.,** Alterations in dihydromorphine binding in cerebral hemispheres of aged male rats, *J. Neurochem.,* 36, 784, 1981.
99. **Hirschhorn, I. D. and Makman, M. H.,** Unpublished studies.

# INDEX

## F

Feeding, see also Diet; Starvation
   effect on lifespan (*C. elegans*), 70—72
   gonadal function and, 168
Fertilization, differentiation of micro- and macronuclei during, 3
Flatworm (phylum), maximum ages of individuals
   in, 124
Freeze-fracturing, 21
FUdR sterilization, of *C. elegans* populations, 65—
   66

## G

Gamma-amino butyric acid (GABA), synthesized by
   leech neurons, 106
Gastrodermal cells, in regressing hydranths, 27
*Gemma gemma*, 158
Genetic analysis, 60—61
Genetics, hermaphroditic mode of inheritance, 49
Genetic studies (nematode)
   Dauer larvae, 81
   isolation of newly induced longevity mutants,
      83—84
   lack of heterosis in, 84—85
   marker stocks, 81—83
   mutation conferring ability to grow in axenic me-
      dia, 81
   variability in lifespan in, 85—89
Glial cells, leech, 106
Glutamate, in CNS of cephalopods, 173
Glycine, in CNS of cephalopods, 174
Gnathobdellid leeches, 97, 111
Gomori histochemical method, modified, 32
Gompertz plots, 68—70, 87, 89
Graft rejection, of earthworms, 127

## H

*Haementeria ghilianii*, 97, 98, 100, 103, 112
Haemopis, 97
*Haliplanella luciae*, 16
Hard clam
   age determination of, 156
   age-height statistics for, 157, 158
   early ontogenetic stages of, 146
   ecology of, 144—146
   laboratory culture of, 144
   life history of, 144—146
   predators of, 145
   shell growth and form, 146—150
   shell microstructure, 150—155
      age determination and, 155
      cyclical microgrowth patterns in, 152
      growth breaks in, 154
   structural features of, 149
   survivorship curves for, 148

"Hatch-off", 63
Hazard rate, calculation of, 89
*Helolodella triserialis*, 97, 98, 100
   embryos of, 104
   embryonic ganglion of, 105
*Hirudo medicinalis*, 97, 98, 112
Histamine, in CNS of cephalopods, 173—174
Histogram, of mean lifespan, 88
Horseradish peroxidase (HRP), 98, 102
*Hydra*, 16
*Hydra attenuata*, single cell types from, 25
Hydranths, 16
   functional capacity of, 26, 28
   gastroderm of
      digestive cells in, 20, 28, 29
      in regressing individuals, 27
   hydrolytic enzymatic activity in, 21—23, 32
   individual organisms, 38
   lifespan of, 36, 37, 40
   protein content of, 34
   survivorship curves for, 17, 20—21
   tentacles of
      ectoderm, 24—26
      progressive alterations in, 19, 24—27
5-Hydroxytryptamine (5HT), in CNS of cephalopods, 173—174
Hypochlorite treatment, to isolate eggs for mass cultures, 62—63
Hypotrichs, clonal aging in, 6

## I

Immortality, 10
Insecta (phylum), maximum ages of individuals in,
   124
Invertebrates
   aging research and, 122—123
   defense and immunity in lower, 123
   histocompatibility reactions in, 132
   $\beta_2$-microglobulin-like molecules in, 134
   studies on senescence and death in, 123—124

## K

Kinetin, 9

## L

Larvae, of *Mercenaria mercenaria*, 145
Leeches
   aging in
      behavioral changes, 112
      morphological changes, 112
      reproduction and, 113
   behavioral characteristics of, 97
   central nervous system of, 99
   development of
      biochemical, 103—105